# Supervision of Technical Staff

# Supervision of Technical Staff

## An Introduction for line supervisors

R. WESTON
*Leicester Polytechnic*

D. C. NORTON
*Formerly at Bromley College of Technology*

M. GRIMSHAW
*North East Surrey College of Technology, Epsom*

ROYAL
SOCIETY OF
CHEMISTRY

**British Library Cataloguing in Publication Data**
Weston, Rodney
  Supervision of technical staff.
  1. Laboratory technicians. Supervision
  I. Title   II. Grimshaw, M.   III. Norton, D.   IIII.
  Royal Society of Chemistry
  502′.8

  ISBN 0-85186-423-6

Published by the Royal Society of Chemistry, Thomas Graham House, Science Park,
Milton Road, Cambridge CB4 4WF.

Phototypeset by Rowland Phototypesetting Ltd
Bury St Edmunds, Suffolk
Printed by St Edmundsbury Press Ltd
Bury St Edmunds, Suffolk

# Preface

Although there are many books available on the general subject of supervisory management, few appear to have been written for what is numerically the largest group of supervisors, the first line supervisor. The majority of publications on supervision contain a considerable amount of material which is more appropriate to the higher levels of management, *e.g.* budgetary control, strategic planning, or specialist areas such as work study. In this book we have concentrated upon the skills necessary at the supervisory level of the organization and specifically those relating to the supervision of staff. Where it has been necessary to consider wider management problems these have been viewed from the position of the supervisor.

Rather than attempt to deal with the general topic of supervision we have produced this book specifically for those with, or seeking to assume, responsibilities for the supervision of technicians in educational, industrial, medical and research establishments. While the actual skills involved are the same as for other areas of supervisory management we felt that it was important for readers to be able to relate the text to their own experience and organization.

This approach has enabled us to be considerably more specific than would be possible in a publication relating to general supervisory skills and will, we hope, be both an aid to learning and to discussion. It has also allowed us to give more space to those aspects of supervision of particular importance to laboratories, *e.g.* training and safety.

We have also been able to include practical examples taken from our own and colleagues' experiences, not to criticize present management standards, but as a means by which the reader can gain from problems faced by others.

The role of the technical supervisor is often underestimated in both industrial and educational laboratories. The competent supervisor provides an important interface between management and the 'production workers' of the laboratory. The fact that these 'workers' may not perceive themselves as such but as 'staff' or 'technicians' makes effective man management particularly important. Changes in ideas and expectations as to supervisory style, with an unwillingness on the part of staff to accept the authoritarian approach, together with an expectation to be more involved in the planning of their own work, has further emphasized the need for adequate training for first-line supervisors.

The need for the supervisor to adopt a flexible role as group leader will be further increased as technological change in the form of micro-processor controlled equipment, robotics and management information systems come together to affect staffing requirements and the supervisory role.

This book is designed to provide an introduction to supervisory skills for:

(a) technicians working at the bench with first line supervisory positions;
(b) more senior technical staff responsible for sections, specialist services or departments;
(c) recent graduates taking scientific posts with responsibilities for technical staff;
(d) full or part-time students of the Business and Technician Education Council (BTEC) units in Supervisory Studies, Laboratory Science and Administration programmes or Continuing Education courses; and
(e) laboratory staff studying at college for National Examination Board Supervisory Studies (NEBSS) qualification, or following the distance learning routes of NEBSS or Laboratory Supervisory Training Services (LSTS).

# Contents

CHAPTER 9
# Training

CHAPTER 10
# Counselling and Discipline

CHAPTER 11
# Industrial Relations: the Supervisor and the Trades Unions

CHAPTER 12
## Health and Safety

CHAPTER 13
## The Law and the Supervisor

CHAPTER 14
## The Supervisor and New Technology

# Supervision of Technical Staff

# CHAPTER 1

# Organization

## ORGANIZATIONAL STRUCTURES

Every employing body, be it a large industrial concern, university department, school or private research laboratory, has to develop an organizational structure to enable it to fulfil its objectives in a planned and co-ordinated manner.

The conventional organizational structure is represented as a pyramid, this being the shape produced by drawing the structure on the basis of the numbers supervised. In practice, of course, this is an over-simplification as each individual within the structure develops formal and informal relationships with others. The latter is particularly important within those laboratories where technicians and senior scientists form the permanent staff, supplemented by a large number of transitory postgraduate and other research workers who may have little experience of the normal working environment.

Before considering the types of relationship within the organization it would be useful to attempt to define the three groups illustrated in Figure 1.1.

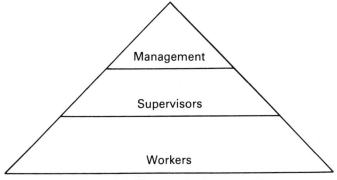

**Figure 1.1** *The organization pyramid.*

**Management**

P. W. Betts in 'Supervisory Studies' (MacDonald and Evans, 1980) states that management implies control by remote or administrative means, indicating certain functions, such as deciding policy and planning, with people reporting to the manager. This is often paraphrased as 'getting work done through others'.

The supervisor may also have responsibility for planning and controlling work, but by close contact rather than more remote means.

**The Supervisor**

As the member of staff who directly manages workers, the supervisor may have authority to engage, fire, reprimand, or discipline, as well as organize work and output. The supervisor under this definition usually organizes or directs the work of others by giving direct instructions, although subordinate supervisors may be involved as an additional layer between supervisor and worker.

**Workers**

The final, and in the normal structure the most numerous group within the organization consists of the workers. These are the people who achieve the objectives of the organization, be this building a product, providing a service, collecting or paying out money *etc*. Within the technical sphere the worker or operator uses technical knowledge manually; the supervisor controls the work of others by using technical knowledge theoretically.

If these definitions are applied to the normal laboratory it will be seen that in the majority of situations even the most senior technical staff may be classified as supervisors rather than managers. Even those who enjoy the title of 'Laboratory Manager' usually manage by direct rather than indirect means. The actual title of a specific post may not be directly related to the functions of the post, as there may be a number of reasons relating to the status of the post-holder which influence the use of job titles rather than the nature of the work done. Conversely the current use of grade numbers, rather than the earlier job titles, within the British university technical service may to some extent reflect upon and reduce the status of the staff to whom they are applied.

Many scientific or academic staff within laboratories are in fact workers rather than managers or supervisors, although they may enjoy a higher status than experienced members of the technical staff in supervisory posts. Indeed, when the above definitions are applied to many educational laboratories/establishments, it is often difficult to identify those who are responsible for carrying out the management functions, which explains a number of the problems encountered later in this book. These difficulties make clear lines of command hard to establish and help to explain why relationships between technical and scientific staff are often not as good as one might like.

## TYPES OF RELATIONSHIP

### Line

The formal direct relationship, based on the Scalar principle, spreads down from the head of the organization through the levels of hierarchy until it reaches the workers at the bottom of the pyramid (Figure 1.2). For the line structure to work effectively, every supervisor/manager must have authority for his staff, with each subordinate being responsible to only one supervisor. Provided this condition is met, then the line structure has the advantage of being easily understood, with each individual being fully aware of his relationship with others in the organization.

Within the line structure given in Figure 1.2 there are four levels.

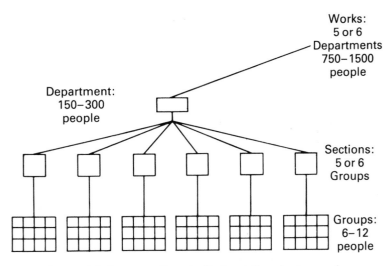

Works:
5 or 6
Departments
750–1500
people

Department:
150–300
people

Sections:
5 or 6
Groups

Groups:
6–12
people

**Figure 1.2** *Structure of a typical industrial/production works.*

The unit, or primary group, consisting of 6–12 people (workers) provides the lowest level, under the control of a supervisor who probably has the title of shift leader or foreman. Five or six of these units are grouped together in a section consisting of 30–60 people. The section head may be referred to as either a supervisor or manager depending on the organization. The third level, consisting of possibly five sections, and some 150–300 people forms a department, with five or six departments comprising the works. In this way each senior member of staff has five or six subordinates reporting to them.

Within the technical structure introduced to British Universities in 1972, a similar line structure appears to have been adopted, provided that all other job groups employed by the university are ignored (Figure 1.3). Organizational structure for a local government organization (college) and a research council, are given in Figures 1.4 and 1.5. In practice, however, the artificial technical grading structures given in Figures 1.3 and 1.4 should not be confused within an organization or reporting structure which would include academic, research and other staff groups—technical staff do not exist in isolation.

## SPAN OF CONTROL

The number of staff supervised by each supervisor gives the 'Span of Control' exercised by that supervisor. Directly related to this span of

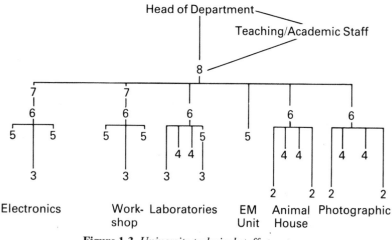

**Figure 1.3** *University technical staff structure.*
(Based upon manual of implementation 1972)

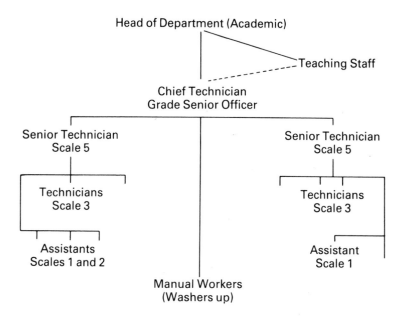

**Figure 1.4** *Typical local government college structure.*

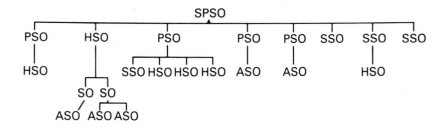

**Figure 1.5** *Organizational structure of a typical research establishment.*

control within the organizational structure is the number of levels within that structure. These two factors taken together not only affect the structure of the hierarchy but also the manner in which it functions.

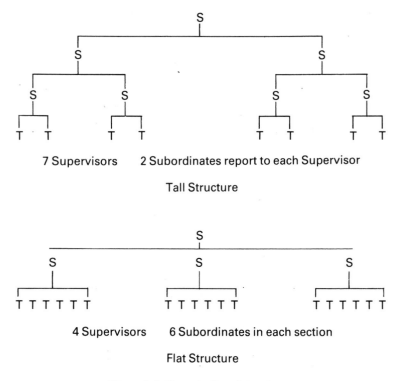

7 Supervisors    2 Subordinates report to each Supervisor

Tall Structure

4 Supervisors    6 Subordinates in each section

Flat Structure

**Figure 1.6** *Organizational structures.*

The 'tall' structure (Figure 1.6) where few subordinates report to each supervisor or manager, gives rise to many levels having narrow spans of control. This is often associated with a centralization of authority which by its very nature tends to be less responsive to change. The small span of control will normally provide less interesting and demanding jobs particularly where the technical supervisor reports to a scientist providing direct control.

The alternative 'flat' structure has few levels with more staff reporting to each supervisor or manager. This type of organization tends to be more flexible, reacting more quickly to change. Within this style of organization decision making is often delegated to the section or unit supervisor who will make the decision on the spot, without reference to higher authority. Consequently, the supervisor benefits from a greater degree of job satisfaction and involvement in the running of the organization.

While the supervisor may thus benefit, subordinate staff could well find themselves being faced with many rivals for promotion and could feel frustrated at the lack of opportunity for advancement to supervisory posts. The alternative 'tall' structure offers greater chance of advancement to supervisory grades, producing more 'chiefs' with relatively few 'indians'.

The structure thus has to meet the needs of the organization, its employees and clients. It is interesting to consider whether these needs have been taken into account in the development of the structures illustrated and whether they do in fact offer the best means of fulfilling those needs.

## RELATIONSHIPS

### Staff Relationships

Staff relationships are those between the manager and his Staff Officer or Personal Assistant (Figure 1.7). In theory this does not introduce an additional level of authority, although in practice this is what often happens as the assistant may assume, or be delegated, some of the functions of the manager. With the conventional personal assistant/manager relationship, where the assistant 'controls' access to the manager, this may not present a problem except where the person seeking access is of insufficient status or confidence to challenge the apparent authority of the personal assistant. On the other hand, staff may approach the personal assistant for information or advice if they do not want to demonstrate their ignorance to the manager.

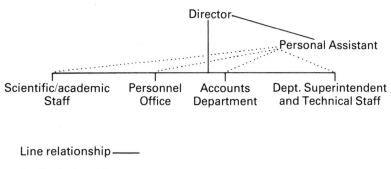

**Figure 1.7** *Line and staff relationships.*

In some educational establishments a staff relationship also exists between the Director or Head of Department and the Laboratory Administrator who may act on behalf of the Head in respect of non-academic staff, routine organization of the building, safety, *etc.*, but not in respect of scientific or academic staff. Under such circumstances the Administrator combines a 'Functional Relationship' with the 'Staff Relationship'.

## Functional Relationship

Every organization employs a range of functional specialists to provide advice and services to managers and supervisors. The Personnel Department has a functional relationship within most organizations and this relationship is on the whole clearly understood and accepted by supervisors. Within recent years two other functional specialists have had an increasing impact on the role of the supervisory technician. These are the Safety Officer and the Training Officer. The precise role of these two will vary between organizations; in most cases they will have an advisory role with no direct authority over supervisors or other technical staff. In some circumstances Safety Officers may in fact have an executive role, in addition to their advisory function, which should be indicated in the employer's Safety Policy or in Codes of Safety Practice.

Normally, the specialist deals through the supervisor rather than directly with subordinate technical staff. Handled correctly, the information and skills provided by the functional staff will enhance the status of the supervisor and increase their effectiveness. Unfortunately, the specialist and supervisor may not fully appreciate their respective roles, with the specialist by-passing the supervisor and the latter thus feeling his or her position undermined. The technical supervisor may also resent the fact that, in their opinion, the functional specialist may not have the technical expertise to provide the correct advice, or feel that they do not understand what is really happening on the ground.

The functional specialists will need to show that they appreciate the supervisors' role and the problems of implementing advice. To this end they should make the effort to establish a constructive and supportive relationship with the supervisor. Given the right attitude on the part of the specialist the supervisor may overcome his or her natural suspicions and come to welcome the fresh insight and support the specialist can supply. But unfortunately there would appear to be an irresistible temptation for such specialists to by-pass the supervisor, either dealing directly with subordinate staff or reporting directly to management.

In the case of the Personnel Section the problem is increased as, in the absence of an effective organizational structure, the Personnel Officer may act to fill a vacuum by developing or assuming executive powers. Such powers may not be confined to the contractual aspects of terms and conditions of employment but may extend into areas which would normally be the responsibility of the line supervisor. This can lead to direct intervention within the group.

**Line By-Pass**

Under certain circumstances line by-pass relationships may in practice be an effective method of obtaining action immediately. Under such circumstances a supervisor faced with a particular situation (*e.g.* safety hazards) by-passes those between himself and the manager responsible for authorizing appropriate action (Figure 1.8). Such by-pass procedures are said to be invaluable in preventing minor situations developing into major problems during the time it would normally take for the matter to move through the line structure.

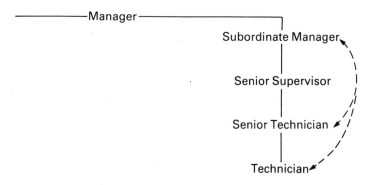

**Figure 1.8** *Line by-pass in which the subordinate manager by-passes the supervisor.*

The regular use of line by-pass procedures can undermine the role of the supervisor particularly if it becomes accepted practice and is allowed to operate as a two-way system. Senior supervisors and management should ensure that the line structure is used as the normal method of communication within the organization. If this proves ineffective it would suggest that the reporting relationships upon which the line structure is based may require streamlining rather than by-passing.

**Informal Relationships**

Formal structures, particularly in large and complex organizations such as scientific establishments, are often too rigid to meet all the needs of the organization even where by-pass relationships exist. A large number of other relationships develop; these may be semi-formal, horizontal relationships between equals in sections which need to co-operate, or of an informal nature in order to avoid bottlenecks and save time.

The point is made by Georgiarle and Orlans in 'Social Skills at Work' (Ed. M. Argyle, Methuen, 1981) that informal groups may work either for or against the organizational goals. In the cases already mentioned they are clearly beneficial but in other circumstances, particularly where there is weak or bad leadership, the informal group may have completely different objectives to those of the organization.

Within the educational sector a number of relationships will also develop between different career groups (*e.g.* technicians and teachers) within sections who may have a reporting relationship but separate organizational structures. For instance, under the Job Evaluation Structure within universities, it is assumed for grading purposes that the technical structure exists in isolation, whereas in fact technicians may have scientific staff, rather than senior graded technicians, supervising them within sections (Figure 1.9). These are not informal relationships but 'reporting relationships'.

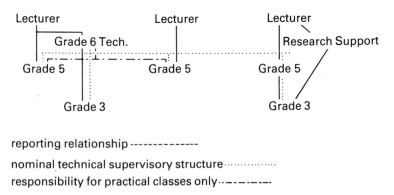

reporting relationship --------------

nominal technical supervisory structure ················

responsibility for practical classes only ··—··—··—··—

**Figure 1.9** *Reporting structure and 'nominal' technical structure.*

An ineffective organization is indicated when the 'organizational' structure and the 'reporting' structure differ or where the reporting relationship differs from the published line relationship. The Chief Technician in different sections may operate line by-pass when dealing with Heads of Department of these sections *e.g.* Life Sciences to Engineering.

In industrial laboratories a more conventional line structure tends to exist with technicians being responsible to a professional scientist as head of section. Line and reporting structures for a local authority college, and industrial laboratory are shown in Figures 1.10–1.12.

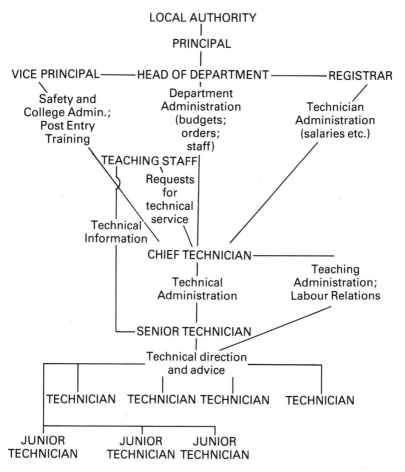

**Figure 1.10** *Reporting structure in a local authority college.*

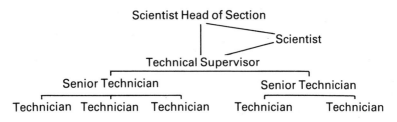

**Figure 1.11** *Line structure in an industrial research laboratory.*

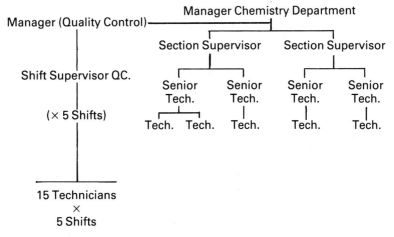

**Figure 1.12** *Line structure in an industrial quality control division. Note the comparatively large number of technicians on shift work under each shift supervisor. The work undertaken by these staff would be of a very routine nature.*

## Technical Structure

While it is convenient to refer to the organizational structure, there is usually not one structure but a number which in theory work together to achieve the object of the organization. The divisions usually occur vertically between groups, *e.g.* manual, staff, management (where the technicians may be in the staff group) or on the basis of occupational groups or classes, *e.g.* scientists, technicians. Within each group there will be separate career progressions and aspirations.

The actual technical structure within employing organizations also varies considerably, it may include a number of grades known by titles other than that of technician, *e.g.* Scientific Officer, Experimental

Officer, Technical Officer or Medical Laboratory Scientific Officer, or it may be completely integrated with that of other laboratory workers.

One of the problems with those organizations which keep technical staff as a separate class, distinct from the scientific, academic or research staff with whom, or for whom they work, is that such a structure is incompatible with a normal line organization. This may result in conflict between the technical supervisor, who has nominal responsibility over a technician, and the scientist to whom the technician actually reports. Alternatively the technical hierarchy is reduced to a nominal structure which comes to exist for grading purposes only. Generally, group conflict is likely to occur within either of these arrangements, with groups defending their boundaries and blaming the other groups for any difficulties which may arise.

Under this type of structure, technical staff who do manage to transfer to the scientific grades may not only fail to be accepted as equals by the existing scientific staff, but often encounter hostility from their former colleagues.

Another problem worth noting is that the aspirations of graduate entrants to the technical grades may differ from those of the traditional technician, and that these may work against the maintenance of an effective service. These problems are also encountered where an academic appoints a recent graduate to a technical grade while regarding them as a research assistant or allowing them to register as an MSc student.

The traditional separation between the scientific and technical classes was based upon the method of qualification, with the scientist following a full time route and the technician studying by day release. However, with the increase in the number of graduate entrants into the technical grades this distinction is no longer applicable.

For a number of years the Scientific Civil Service has operated a staff structure, without distinction between classes, based upon five grades. In practice, educational hurdles make progress above a certain level difficult for the HNC holder. The Institute of Medical Laboratory Sciences has published proposals for a unified grading structure within the National Health Service Laboratories, but these have not been adopted by the NHS management. The argument for combining the present laboratory scientific officer grade (technicians) and other scientific officers (mainly biochemists) is based on increased efficiency and improved morale.

The educational sector has retained the traditional separation between technicians and other laboratory staff with a nine grade technical service plus pseudo-academic grades and professional scientific staff. As might be anticipated this produces some difficulties associated with intergroup conflict and the absence of a conventional line structure. In the 'Istox Laboratory Management Survey' (Oxford Branch I.S.T. 1979) only 1% of chief technicians within the university structure were recognized as having sole responsibility for controlling staff.

## INTERGROUP RELATIONSHIPS

The effect of intergroup relations upon the performance and motivation of the group may be considerable and, as such, the maintenance of good relations is of importance to the supervisor; unfortunately many aspects of the situation are, by definition, beyond the supervisor's direct control.

The term 'Intergroup' is often applied to the interaction between separate units, *e.g.* marketing and production. Within laboratories, conflict may arise between established and grant funded units, between research and teaching requirements (particularly where group membership overlaps) or between different career groups, *e.g.* technicians or academics/scientists.

There are a large number of possible causes of poor intergroup relations (Figure 1.13) primarily:

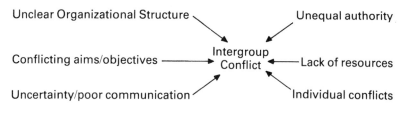

**Figure 1.13** *Causes of poor intergroup relationships.*

(1) *Unclear organizational structure*
This is a common problem in many educational and research organizations where undue emphasis is placed upon academic freedom or where the management function is neglected. The problem also arises in hospitals, where medical staff can order expensive treatments without worrying about budgets which are allocated to supporting sections, *e.g.* X-ray, nuclear medicine *etc.* In response to this situation

the supervisor will need to establish effective informal working relationships to minimise the lack of formal structure.

(2) *Conflicting aims and objectives*
This problem is also associated with the absence of an effective management structure and lack of co-ordination between groups, units and types of staff.

(3) *Uncertainty*
This may result from the above, from poor planning, failure of communication *etc.*

(4) *Unequal authority, support or power*
This may take one of two main forms. The first arises where one working group or unit has a greater degree of support from those in authority, *e.g.* research being favoured over quality control, or the Head of Department appearing to be interested in his own group rather than all staff.

The second form is found where one work or career group is dependent upon another for a number of essential activities, *e.g.* technicians upon personnel staff for hiring, firing and discipline.

(5) *Lack of Resources*
This may arise out of lack of support for particular groups, or from certain groups being thought to be overfunded in comparison with others. The overfunding may be real or as a result of differences in the expenditure policy of the section head, *e.g.* with one section being committed to providing technical staff with the best possible equipment, compared with another spending its funds on items not related to the work of the technician.

The problems of lack of resources are assuming greater importance with the current expenditure cuts and redundancies.

(6) *Personnel*
Difficulties resulting from differences or arguments between individual members of staff in different groups arise out of work related matters or outside factors. In the current economic conditions individual conflict may also arise from possible redundancies, lack of continuation of employment, job sharing, *etc.*, which affect some groups more than others rather than being evenly distributed throughout the organization.

It is unrealistic for the supervisor to pretend that intergroup problems will not occur, or that once they have arisen they can be ignored and will eventually go away.

The supervisor should learn to recognize the problem and identify the causes as it may also be possible for the skilled supervisor to turn the situation to the advantage of the group. Whilst conflict and disagreement may be disruptive, it can also be used to focus attention on a problem or as a unifying force if the unit can be bonded together in competition against other groups.

An example of this might be a group responsible for maintenance of buildings who control their own resources and will provide the necessary skills to carry out certain work. The laboratory supervisor finds that jobs do not get carried out or take excessive time. By using this maintenance group as an example the supervisor can highlight the problem to a management group who are in a position to alter the system, and also unify his own group by showing his concern over problems which directly affect them.

Where the supervisor finds that the intergroup conflict cannot be used constructively it will be necessary to consider the removal or mitigation of the problem. As there is little point in the supervisor allowing the group to continue with a conflict that it can not win, this may require the changing of methods of the group, increasing results *etc*. If, on the other hand, the changes necessary to reduce the conflict would result in stress or demotivation within the supervisor's own group, there would be a case for attempting to persuade the other group to change in order to reduce the conflict.

Where it is not possible for the cause of conflict to be removed or reduced, the supervisor will need to minimize the detrimental effect this has upon staff motivation. This may be attempted by the use of a participative style of management and by involving all members of the group in discussion of the problem. Even if this does not produce a solution it will provide a means by which staff can give vent to their frustration and this can then be managed or controlled by counselling individual members of staff, together with team building and bonding techniques. Care must be taken to ensure that staff who cannot achieve their personal objectives as a result of intergroup conflict, do not become cynical about their lack of ability to change things.

The supervisor may also attempt to reduce the apparent size of the problems caused by intergroup activities by giving emphasis to other goals of the group in an attempt to reduce the impact of the conflict.

# CHAPTER 2

# The Role of the Supervisor Within the Laboratory

## STATUS OF SUPERVISORS

The precise functions and responsibilities of the laboratory supervisor will vary considerably, depending upon the type of organizational structure within which they operate and the effectiveness of management.

In the first chapter, we established that the difference between managers and supervisors was normally considered to be one of distance or closeness of control. However, it should not be forgotten that the supervisor–subordinate relationship exists at all levels within the organization, from the most senior director to the technician at the bench.

Supervision therefore occurs throughout the organization, which may be considered as being based upon a supervisory team at all levels. Despite this, only the lower grades exercising supervisory functions are classified as supervisors with the higher levels being awarded the title of management. The ratio of planning, problem solving, and decision making within the job has been offered by some authors as further means of distinguishing between management and supervision. However, even where technicians perform these functions as Laboratory Superintendents, they are frequently considered to be not managers but supervisors.

In reality, the crucial factor appears to be not job content but status, with similar jobs in different types of organizations being classed as supervisory or managerial depending upon the regard in which the employer holds them (Figure 2.1).

The status of supervisors may be indicated by symbols of rank conferred by management, such as an office, desk, telephone, secretary *etc.*, by the use made of the supervisors as a channel for

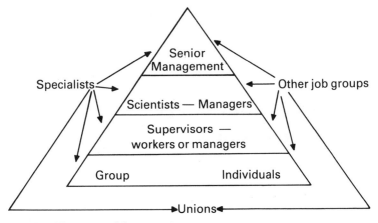

**Figure 2.1** *The status of the supervisor.*
Adapted from the 'Burton Manor Supervisory Management Handbook'
(Newman Neame 1953)

communication and the extent to which they are consulted on matters
affecting their section. The status conferred upon the supervisory
technician may also be indicated by the use of job titles. The use of
grade names such as Laboratory Manager, Chief Scientific Officer
or Laboratory Superintendent confer a certain measure of respect or
authority upon the holders of the particular post, whereas the use of
grade numbers, *e.g.* grade 7 technician or T8, are either meaningless to
the uninitiated or may indicate a lack of status. No doubt it is for this
reason that many technicians within 'number-based' structures invent
titles for themselves.

The views of subordinate technical staff may differ from those
of management. In industry and many large educational laboratories
it is not unusual for operatives or junior technical staff to regard
their supervisors (or foremen) as part of management while the
organization may view them as part of the workforce.

In practice, it might be felt that supervisory technicians are in an
impossible situation being neither manager nor worker, by-passed by
functional specialists, with their status often undermined by lack of
support from senior management, and their role not recognized by
scientists. There are pressures from above and the supervisor is
expected to achieve objectives set by management, whilst those
supervised will expect the supervisor to protect them from the actions
of management. Rarely can this role conflict and these different group
aspirations be avoided.

This question of status is important as the attitudes of all members of staff are shaped by the way in which they are treated by their seniors. Within laboratories the problem is exacerbated by the fact that few senior staff are trained in supervisory or managerial skills and this is often reflected in the style of management adopted. In many cases it will be appreciated that although the technical supervisor has a formal position within the organization this tends to be ill-defined, depending considerably on the informal acceptance by the group or unit. This acceptance may, in turn, be based more on the supervisor's technical skills and knowledge, coupled with leadership style, rather than the status granted by the organization.

A number of factors will affect the operation of the supervisor and the performance of the group. One of the most common is intergroup conflict, discussed above. It is mentioned here, as one of the causes of such conflict may be the status of the supervisory technician and the degree of formal control the supervisor is allowed. While it is often difficult to extend this formal authority the supervisor may increase his status, control, and influence by the use of informal authority.

**Expanding Control**

The individual supervisor should be advised to enquire tactfully as to the methods of extending formal control such as committee membership, involvement with management in forward planning, *etc.* Where progress is unsatisfactory there are informal methods which may be used, such as:

(1) *Expertise.* The supervisor may use his technical skills and knowledge to build up a reputation as a person with expertise, either in a specialist field or for general techniques. This expertise may be expanded into areas that will be in demand by other groups or sections, *e.g.* computing, safety, training, college courses *etc.*, and they will come to recognize the skills offered.

(2) *Favours.* The supervisor may use favours done for colleagues and people in other groups as a means of increasing influence. The use of such methods needs to be handled with considerable subtlety, for success depends upon people not recognizing that the supervisor is attempting to increase his influence. There are dangers in the use of favours, for others may take advantage of willingness to help, or the supervisor may be seen as 'crawling' to the management, or of setting a

precedent which will affect the future activities of both the supervisor and their group.

(3) *Self-Advertising.* Status may be acquired within the organization through duties not directly related to the post in which one is employed. Typical examples of this form of self-advertising would be by holding office in a professional body or Trade Union, although involvement in the latter may be a double edged weapon, as the supervisor may be associated in the view of some managers with 'left-wing' or anti-organizational activities.

Once established as a recognizable figure outside the organization the supervisor may be considered as an individual with superiors being prepared to listen to his ideas.

(4) *Developing Support.* The effective supervisor intent upon increasing his influence will develop support among fellow supervisors and work colleagues by lobbying and discussion, ensuring that his views are known on topics requiring action.

## LEVELS OF SUPERVISION

While the role of the supervisor within any structure varies considerably with the nature of the establishment, the work load, and interaction between the groups within the organization, it is normally possible to identify a number of levels of supervision applicable to both laboratories and other organizations. A number of these were mentioned briefly when discussing organizational structure in Chapter 1.

### Sub-Primary Group

These are under the control of technical staff working alongside their colleagues at the bench. These 'bench supervisors' may not have direct control over staff but are responsible for ensuring tasks are accomplished. In addition they may provide training and instruction for junior staff on a 1:1 or 1:2 basis. The sub-group supervisor may be a grade 4 or 5 technician within the educational services or a scientific officer in the civil service establishment.

### Primary Group

The primary group is the basic supervisory unit shown in Figure 1.2 as consisting of 6–12 people. Within scientific establishments this may be a research group, a functional group of laboratories, *e.g.* microbiology laboratories, or service section such as quality control.

## Supervisors

In industry the supervisors of these groups form an important link between operators and management, with a similar role being performed in laboratories where the supervisor and subordinate are drawn from a similar background. When the supervisor is a research scientist or academic of considerable standing in the organizational hierarchy there may be problems associated with the scientist not recognizing his function as a primary group or line supervisor. Many scientists in this situation not only do not recognize their supervisor function but only want to be concerned with academic matters not wishing to 'dirty' their hands with administrative and personnel problems. These people are also often the ones who are most reluctant to grant status to the laboratory supervisor. Perhaps in part this is because of the disparity of qualifications and background and it will be interesting to see how the situation develops when the graduates now entering the technical grades progress to senior positions.

In those primary groups where the organization does not appoint a leader, an unofficial one may develop. A similar situation may arise in larger groups which split into unofficial small groups. This natural tendency for a leadership vacuum to be filled unofficially may cause difficulties for the organization, for rather than being part of the organization, supported and guided by it, the leader will have developed from, and reflect the view of, the workers. Where the official organizational structure is different from the actual line structure of the establishment, the resultant absence of supervisory status or recognition for those technicians in intermediate grades may result in them assuming the mantle of unofficial leaders either within the group or through involvement in trade union or similar activities. From the organizational point of view this may not be desirable and points to the need for an effective organizational and communicative system.

## Section Supervisor

The section normally forms the second level of supervision within an organization with the section supervisor being responsible for about six primary groups totalling 36–72 people. Within many scientific establishments the position is complicated by the word 'section' being used to describe a discrete primary group often of a service nature, *e.g.* electronics or electron microscopy section. Based upon the numbers of staff, the term might be more appropriately applied to a small university or college department or a hospital pathology department rather than the individual units.

The role of the section supervisor is normally concerned with the allocation of duties, co-ordination of the primary groups and the day to day affairs. The Chief Technician or Departmental Superintendent may be considered to fulfil the role of a section supervisor even though he or she may operate on a departmental basis.

**Departmental Supervision**

In the industrial situation the department may consist of six sections, or 150–300 people, with the departmental supervisor being concerned with planning and costing. A certain amount of policy making will also take place at this level, *e.g.* forecasting, budgeting, *etc.*

**Work Supervisor**

This is the senior supervisor in industry responsible for 6–8 departments. These last two levels of supervision are more concerned with management than the supervisory function and within this book we shall be concentrating on the functions of those exercising authority at the group or section level.

## FUNCTIONS OF THE SUPERVISOR

All supervisors, no matter what the purpose or structure of the organization or the level at which they exercise their supervisory function, will undertake certain common activities. Obviously the extent and ratio of these activities will differ between posts and levels.

**Specialist or Technical Skills**

These will relate to the work undertaken by the group, section or department. In the case of technical supervisors they will normally be the bench skills in laboratory techniques, electronics, engineering *etc.* in which the technician was involved prior to promotion. Naturally this body of knowledge will need updating while the person is in the supervisory position if the respect of subordinates is to be maintained.

There will almost certainly come a time when the supervisory duties take precedence over these technical skills and the supervisor will come to rely upon man management skills to provide him with a means of obtaining the best technical advice via his subordinate supervisors.

## Planning

Overall planning is a function of management by which objectives for the supervisor's groups are set. Within these objectives the supervisor will be responsible for setting the detailed objectives for the group and individuals within it for the day, month or year. Often pressure of work is such that the group operates only on a day-to-day basis and although longer term targets are borne in mind there may be little forward planning as to how best to achieve them.

Having decided upon the objectives the supervisor will need to consult and agree with subordinates how and when the necessary action will take place together with the resources required. At this stage it may be necessary to communicate with superiors as to allocation of resources to meet the overall objectives, and colleagues in other groups where co-ordination is necessary.

## Organizing

Within the planning function the organizing role of the supervisor has already become apparent, for even at that early stage it is necessary to consult with staff to ensure that the plan being formulated is capable of implementation. At this and subsequent stages the supervisor will require to use communication and motivational skills to organize the work of the group. In organizing the work within the unit the supervisor will assign tasks to those technicians under his control, he will co-ordinate the work and to some extent the relationships between members of staff. In this context the supervisor has:

- *Responsibility*—which may be defined as the area in which end results must be achieved.

- *Authority*—the right not only to control, within limits, others but more importantly to use resources. In fact staff may be looked upon as one of the resources of an organization together with money, equipment *etc*. This implies that authority can be delegated downwards from management to supervisors. In theory this is true, but in practice many managers are reluctant to delegate full authority.

- *Accountability*—the need to produce, and be judged upon, results for areas of responsibility within the limits of the resources made available.

The supervisor, in organizing the work of staff, should provide a clear line structure within the group, with each individual receiving orders from only one person. Unfortunately, in many cases the supervisor will not have control over the structure, or family tree, of their section. This may be based on Job Descriptions and a structure established by senior management or on the basis of a management/trades union grading structure, agreed separately for each job group within the unit and may have little similarity to the structure the supervisor feels is necessary to achieve maximum results.

**Controlling**

Although considered as a distinct function, controlling is, in practice, an integral part of the supervisors organizing and planning roles. Control is concerned with ensuring that what is planned actually occurs and with correcting errors or mistakes before they become serious. For this to be effective it obviously needs feedback at all stages of the process.

As indicated above, control may be applied to personnel resources (money and equipment) and technical procedures. While the actual methods of control will vary, each will need to contain three basic elements (Figure 2.2):

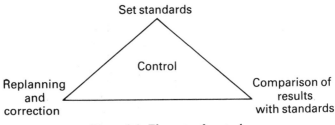

**Figure 2.2** *Elements of control.*

- Setting the desired standards of performance or results; deciding on a system of measurement.
- Comparing the results achieved with the targets set.
- Deciding on and taking corrective action, *e.g.* replanning.

## Job Evaluation

Job evaluation forms part of the supervisors 'control' activities but is selected for particular attention because of its importance as the basis of many technical structures. Information may on occasions be required from Supervisory Technicians on the details of the job under investigation. At this stage influence can be used to produce a structure which will fulfil the planned needs of the technical service. Details of how job evaluation is carried out are given later in this book.

The importance of using the structure as a channel for communication has been mentioned, unfortunately, it is often the case in practice that supervisors in their hurry to obtain results may themselves by-pass the line structure and undermine the system of work they have instigated. Job descriptions may be used to the advantage of the supervisor in organizing the work in the section providing that they are relatively specific in detailing the functions of individual members of staff and are not over generalized 'bench marks'. The latter are used to grade staff but provide little information as to their precise duties or how these dovetail with those of other staff (Job Descriptions are considered in more detail in the chapter on training). The use of an organizational diagram for the group showing lines of authority and responsibility based upon job descriptions and the views of staff may prove useful in avoiding misunderstandings, particularly on those occasions when the supervisor is absent.

## Communication

It can be seen, therefore, that the main functions of supervisors are based to a large extent on communication skills, with subordinates, supervisors, and colleagues, and motivational skills both in respect of self and subordinates. Communication skills are discussed in more detail in Chapter 4.

Communication may be between supervisors and subordinates, intermediate supervisors, or supervisor and manager. In most cases the first two will be by verbal means while the latter may often also involve written reports.

The provision of information is an aspect of communicational skills that particularly involves the translation of company policy into terms relevant to the group, and providing information to members of the group relating to such matters as their careers, or grading.

Skills in human relations will also be required for such activities as selection, recruitment, training, disciplining and motivating of staff.

# CHAPTER 3

# Leadership

Effective leadership will depend upon a number of factors: how people are treated, on the understanding of people, expertise, qualifications, and the support for staff development, *etc*.

Not least of these skills is the issuing and receiving of orders or instructions. These are an accepted part of employment; the supervisor must demonstrate the ability to both give orders and of deciding how and when subordinates will comply with them. The supervisor needs to maintain control of the group, to assess priorities, direct staff, ensure effective working relationships and achieve the desired results for the group.

Effective command within the technical service, where authority and organizational support may be lacking, will depend upon the co-operation of subordinate staff which in turn may depend upon the style of leadership adopted by the supervisor. The supervisor would be in a better position to establish an effective working relationship with subordinates if he was responsible for, or at the very least directly involved in, their selection and promotion. Unfortunately this responsibility often rests elsewhere in respect of selection while the introduction of job evaluated grading systems may have lessened the impact of the supervisor in respect of the promotion of subordinate staff.

In the subsequent sections of this chapter we will deal with leadership and leadership style as it applies to the technical supervisor for whom this book is intended. However, it must not be forgotten that the style adopted by senior management, specialist advisors, and scientists can have a major impact upon the morale of the group. Good team spirit is essential for efficient working but the fostering of a team approach depends not only upon the personal relationship that exists between supervisor and supervised but on such factors as contractual conditions and perks, for example, direct access to an 'outside' telephone, additional holiday, medical insurance, different pension scheme, the use of a company car, *etc.*, which will also influence the development of a team approach towards achieving the objectives of the organization. Such a spirit is difficult where conditions of employment are as disparate as those which exist between management and others within some laboratories, *e.g.* academic and technical staff within the university sector. Not only do members of the academic staff have better salaries but they may benefit from the absence of fixed working hours, college facilities, sabbatical leave, frequent trips abroad to conferences (which often seem to coincide with sunshine or skiing) and better pensions. The technician on the other hand will often find that the academic supervisor does not feel the need to lead by example in respect of both the work and conditions of employment. Members of the academic staff in colleges of education do not enjoy the same freedom as their colleagues in universities and the gap between academic and other staff may not seem as wide, but it is still there.

It might be argued that there is little purpose to be served in the group supervisor working to produce an effective and happy unit if an unsympathetic or authoritarian management is going to destroy their efforts. The need for recognition and rewards for good service which are essential for the morale of the group must be seen to pass down from the managerial or senior scientific level and not just from the line supervisor. It is only too easy to find examples of poor or weak leadership at all levels within scientific organizations, often as a result of the different goals of senior staff and the absence of a clear line structure with recognized levels of authority. This situation is enhanced by a lack of formal management/supervisor training at all levels. Whilst one may sympathize with the supervisor in these situations, the supervisor as a team leader must set standards and, even if not supported from above, attempt to maintain professional attitudes to the work of the organization.

## AUTHORITY

In theory authority is conferred upon the supervisor as part of the mantle of the post. However, in many laboratories there may be little difference in the authority between first line supervisors and bench-workers in the career grades. In practice, at this 'primary group' level, much of the supervisor's authority derives from personal attributes. Factors influencing such authority will relate to the respect the supervisor earns from subordinates by demonstrating skills and competence in his work, including technical expertise. The need to earn the respect of subordinates is often forgotten, with some supervisors feeling that the respect of staff automatically goes with the post. This also relates back to the previous section; the need for the laboratory supervisor to be given authority by the academic or scientific staff. However, all supervisors, whether supported by management or not, have to establish and maintain their credibility.

In many laboratories the working atmosphere is considerably more relaxed than in other areas of employment, particularly in respect of time-keeping and other matters of discipline. This informal style must not obscure the fact that supervisors need to impose discipline in certain areas, such as safety. This can present leadership problems especially if the senior scientific staff are not prepared to support the technical supervisor in enforcing safety regulations to the full.

The problem of lack of support or back-up for technical supervisors is a recurring one within scientific establishments and is of considerable importance. Fiedler ('A Theory of Leadership Effectiveness.' McGraw Hill Book Co., 1967) gives the authority of the leader and the support of the organization as major factors in leadership, together with good leader–subordinate relationships and clear objectives.

The lack of support also extends to the provision of supervisory skills training for both technicians and scientists. The need for such training is widely recognized and the Institute of Science Technology is now offering a postHNC programme, 'Supervision and Management', in conjunction with the Business and Technician Education Council (BTEC), specifically for staff in laboratories.

Most technical supervisors have a large number of diverse demands upon their time with untrained supervisors responding by demonstrating a number of traits. These may include disorganized work, jumping from one task to the next without completing the first, and responding to problems as they arise rather than planning to meet them. Such staff may end up doing tasks which should have been delegated to subordinates or left for a later date while urgent tasks are left until the last minute.

The problem of a diverse and incessant workload is not confined to supervisors in laboratories. In one survey of industry, F. J. Jasinski and R. H. Guest ('Redesigning the Supervisors Job' *Factory Management Review*, **115**, No. 12) showed that the supervisor may be faced with a different task every 30 seconds, *i.e.* 876 per day! This high figure demonstrates the need for supervisors to be aware of the needs of the job and to be able to assess priorities.

## NEEDS

Argyle ('The Social Psychology of Work.' Penguin Books, 1983) identified three needs which the leader must consider and balance to achieve the best results. These needs and their interaction will also influence the style of leadership adopted by the supervisor (Figure 3.1).

The leader is appointed to fulfil the task needs and the objectives of the organization but these cannot be considered in isolation; if the objectives are not reached this affects both the performance and the morale of the individuals concerned and the group.

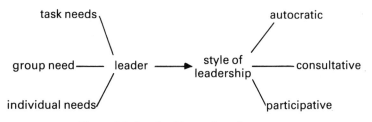

**Figure 3.1** *Leadership needs and response.*

## Task Needs

The formal group will have been established by the organization to achieve certain tasks which may be of a specific or general nature. Adair, ('Effective Leadership' Pan Books, 1983) makes the point that in formal groups the leader is appointed by those in authority, whereas the leader of informal groups with self-chosen tasks is usually elected by the group. Where this is the case, the formal group leader is primarily responsible to those in authority, and only secondarily accountable to the group. Within this framework it is the responsibility of the leader to establish a sense of direction and momentum by deciding, alone or in consultation with the group, upon specific objectives for the group and the individuals within it.

These task needs will exist at all levels of leadership whether the supervisor is responsible for a small section or a whole department. Within a section or single laboratory the objectives may be precise and clearly recognized, *e.g.* carrying out $X$ number of laboratory tests. At higher levels the situation may be complicated by conflicting aims and objectives facing the leader with the need to analyse the problems, balance alternative solutions, and provide constant encouragement to the group.

**Group Needs**

The technician's group or unit will have a number of needs from their supervisor, as has the supervisory technician from his manager. These needs will relate to such matters as the support and organization of the group, recognition or rewards for their efforts, and discipline.

**Individual Needs**

The needs of the individual are complex and the priorities of the various needs will vary with individuals. The leader will need to protect individual members of the group and not 'pass the buck downwards', provide goals for members of the group to enable them to obtain a degree of fulfillment, and possibly act as a 'father figure'.

**STYLE OF LEADERSHIP**

The style of leadership adopted by the supervisor should be flexible and capable of being changed to meet the needs of a given situation. Perhaps the major consideration in the choice of style is the attitude and maturity of subordinates.

Where staff are mature and well-motivated, responsibility for tasks can be delegated with checks through indirect mechanisms (Figure 3.2). At the other extreme, immature or irresponsible staff will require a more authoritarian style—the 'telling' approach. Between these two groups we have the 'selling' approach or persuasive style, and the consultative or 'participatory' style which allows staff a varying degree of input but stops short of complete delegation.

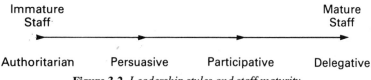

**Figure 3.2** *Leadership styles and staff maturity.*

There may be a need for the supervisor to balance these styles, adopting the one best suited to the staff and the task. It is interesting to note that there is evidence to suggest that where supervisors have been deliberately adopting a style they revert to their natural style in a crisis. Supervisors who have shown sensitive styles of leadership in normal routine situations may find that in the event of a crisis, *e.g.* shortage of staff due to sickness, or extra work at short notice, their staff rally round to be of extra help. Staff whose attitudes may be to carry out work to the letter and no more, may suddenly offer help without any request from the supervisor, simply because they respect the work that the supervisor has done on their behalf in the past. When the whole group is 'threatened', it becomes protective to itself in a kind of 'wartime' spirit. Conversely the supervisor who adopts the uncaring approach is likely to find that their staff do not 'rally round' when things go wrong.

### Authoritarian or Autocratic Style

This was once considered the normal style of management but under conventional situations is now no longer acceptable to subordinate staff.

The autocratic leader sets his own standards, makes his own plans which are usually only given to group members on an incomplete or 'need-to-know' basis and gives orders expecting them to be followed.

The authoritarian leader may keep himself separate from the group both physically, taking tea or coffee breaks with fellow leaders rather than the group, or by the use of titles as a means of distancing himself. Linked with this is the view that he knows what is best, not only for the group, but for individual members of the group, and he may oppose any ideas or initiatives which come from others. As a concession to modern management, such managers may now accept the need for communication between managers and subordinates but do not accept the need to change their style or views as the result of such communication.

The autocratic leader may also operate an organizational structure so that they and only they, get the whole picture as to group progress and activities (Figure 3.3).

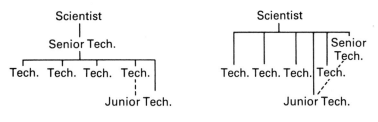

Organizational Structure                    Reporting Structure

**Figure 3.3** *Structures under authoritarian style of leadership.*

As a result of this style of leadership the group will either become very dependent upon the leader or take a passive uninvolved role in the work of the group, producing results only when supervised and being disinclined to provide more than the minimum effort. It is also not unlikely that individual members of the group or sections within it will feel considerable dissatisfaction and little respect for the leader. This style deals with mistakes and matters that have gone wrong in a somewhat military way, punishment or a dressing down being a typical response. Under these circumstances junior staff will do sufficient to avoid punishment but will not develop a professional attitude.

**Participative Style**

The participative style of leadership involves the group in policy making, encourages the group and individuals to organize their own work within the objectives of the group, and seeks to help staff meet their own objectives.

Under such a style the group develops strong cohesive links with a low dependency upon the leader. Group output tends to be high, and not directly linked to the degree of supervision, as staff are working to a high degree of personal satisfaction. The development of a professional attitude in staff is encouraged, and this enables the manager to provide greater career development for subordinates. There is a danger in this style, in that the status of the subordinate supervisors (the senior technicians in Figure 3.4) may be undermined as they may be by-passed in the participative system. This danger is inherent in systems which encourage staff to develop professionally, become more expert and take additional responsibilities.

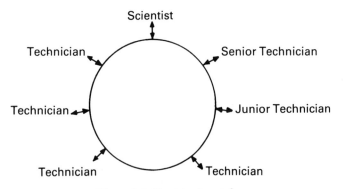

**Figure 3.4** *Participative style.*

## Consultative Style

The consultative style occupies the middle ground with the leader consulting staff and taking into account their view when making decisions. Success depends upon the manner of consultation and the degree to which the advice is taken. This style would seem to be particularly suited to the laboratory situation where there is a need to work through subordinate supervisors and support their position, *e.g.* where technicians provide several service functions.

## Laissez-Faire

A fourth style of supervision only too common in many organizations within the public sector, is that of the abdication of leadership; the *Laissez-Faire* approach. The leader may adopt this style either through an over enthusiastic commitment to the democratic style or through disinterest in the day-to-day affairs of the group. Within laboratories this may occur as a result of an overiding concern with a personal research project or with the leader's own particular section rather than the group as a whole. In such cases discontent will develop within the groups which may split into rival sub-groups concerned with their own limited objectives. Within such systems output within sub-groups may be maintained at average levels provided cooperation is not required between the groups.

Under other styles of supervision rivalry between subgroups, if carefully and subtly controlled, can be productive rather than counter-productive. Each group in its attempts to attract attention from the supervisor may develop the given tasks beyond the normal objective. For example, research groups may try new techniques or attempt to improve their level of expertise by practising a technique, and then use this increased expertise to improve the work output. Teaching groups may develop practical exercises for students in order to provide better technical methods which the academics may use for class teaching.

The emphasis at all times so far as the supervisor is concerned, is to control the extra enthusiasm generated by this rivalry without the subgroup realizing the depth of this control.

## IDENTIFICATION OF SUPERVISORY STYLE

A knowledge of the different types of supervisory style and the ability to identify the styles of others is useful, but of greater value is the ability to recognize and if necessary adjust ones own approach.

The exercise below, modified from one used by a motor manufacturer is designed to assist the supervisor to achieve the first of these
objectives either as a 'self' test or for testing subordinate supervisors.

Record your perception of your leadership style as authoritarian,
consultative or participative.

● **Part I** Now read the two descriptions of supervisory style given
below and mark on the scale the position between X and Y which you
feel reflects your own attitudes.

Attitude X
The average technician has an inherent dislike of work and because
of this needs to be closely supervised, directed and disciplined in
order to get a fair day's work. In practice they do not like responsibility and prefer to be directed in their tasks.
Attitude Y
Work is natural and most technicians are committed to their work
and have a desire to be involved and seek wider responsibilities.
They work best using self-discipline without direct supervision.

X_____Y

● **Part 2** Having marked your position on the scale above, record
your attitude to the five items below.

|  | A<br>Normal<br>behaviour | B<br>Occasional<br>behaviour | C<br>Seldom or<br>never |
|---|---|---|---|
| (1) Detailed instruction should be issued for each task with regular checks on the standards of work. |  |  |  |
| (2) Staff should set their own standards and prepare their own work schedules. |  |  |  |
| (3) Subordinates should be encouraged to assume additional responsibilities even if this means that they become eligible for a higher grading. |  |  |  |

(4) Junior technicians need
pushing to achieve results and to
ensure they meet the required
targets.

(5) The supervisor should
regularly discuss worries and
problems concerning both work
and the organization with staff.

● **Part 3** Score the following points for each answer:

(1) **a1, b2, c3,**
(2) **a3, b2, c1**
(3) **a3, b2, c1**
(4) **a1, b2, c3**
(5) **a3, b2, c1.**

*e.g.* Question 1, level b (occasional behaviour) would score 2 points.

Transfer your score to the scale below:

Authoritarian Style                    Participative Style

0_____5_____10_____15

Compare your position on the scale with your response to Part 1.

## ORDERS

At the beginning of this chapter we stated that the style of leadership adopted by the supervisor will determine the extent to which orders are given and the nature of those orders. With the participative style, orders, either direct or in the form of requests, may be unnecessary as the objectives of the group and any problems that arise will be discussed with all members and a common strategy agreed. There are three main classes of order.

### Open Orders

Open orders or objectives allow the subordinate to decide how they will be achieved. With suitably trained staff this method can be very rewarding to the subordinates. The supervisor in such circumstances needs to show sufficient interest in the methods followed to provide support for their staff without being seen to interfere. The ability to stand back and allow subordinates to develop their own methods and procedures, particularly when these differ from those which the supervisor would adopt, is an important supervisory skill.

## Requests

These may range from mild hints or suggestions such as 'Some of the labels in the chemical store are difficult to read', 'There seems to be some glass in the waste paper basket', 'Would you get the chemical store labels replaced?' or 'Could you remind people not to put glass in the paper bins?' The precise form of the request needs to be carefully measured to match the person for whom it is intended, a hint or mild suggestion being appropriate to a well-motivated member of staff while a direct request may be more suited to others.

## The Direct Order or Command

In general, the direct order or command tends to belong to a more authoritarian age and, as with the autocratic style of leadership with which it is associated, may be unpopular with subordinates. While the use of this approach may antagonize well-motivated staff it may offer the only means by which disinterested subordinates can be made to perform the necessary functions.

If a supervisor finds that he or she is continually resorting to the use of direct orders it is an indication that all is not well within the group and such factors as motivation, recruitment policy *etc.* could benefit from examination.

The direct order still has a place within the repertoire of the leader and that is when an instant response is required in an emergency. Every supervisor will find that at some time a direct order has to be given, even though this may go against the way in which the supervisor would wish to act.

# CHAPTER 4

# Organization, Planning, and the Technical Supervisor

The need for planning and forecasting are readily accepted in manufacturing and industrial laboratories. This concept, however, is less readily accepted as being appropriate to the technical supervisor within the public sector, where the role of such technicians may be seen more in terms of benchwork and routine tasks which require some day-to-day organization but little planning in managerial terms. However, such procedures require considerable planning if they are to be completed within the resources provided in terms of material, equipment, finance, staff availability and the need to coordinate with other groups.

## LEVELS OF PLANNING

### Short-term Planning

Planning of this nature is concerned with the day to day operation of a section, involving the actual preparation and performance of the work, work rotas *etc.*, and forms the lowest level of formal planning within the organization. It tends to be concerned with detail whereas the higher levels involve less detail and more of an overview. Within laboratories, this level involves technical supervisors, scientists and lower level administrators.

### Middle-term Planning

Senior scientific, administrative staff, and some senior technical staff will be involved in middle level planning covering a period of three to five years ahead. This is perhaps easiest appreciated where fixed research grants covering these periods are concerned but in practice the procedure occurs in all circumstances.

**Long-term Planning**

Planning at this highest level normally covers a period of up to ten years and takes the form of general objectives. Within the context of this book only that level appropriate to supervisors will be considered.

## THE TECHNICAL SUPERVISOR'S ROLE IN PLANNING

Planning cannot be considered in isolation nor can a newly appointed supervisor hope for a clean start. In the majority of cases the supervisor will be faced with an ongoing situation and will want to plan for better performance within that situation. The supervisor will need initially to identify problem areas within the group and continually monitor the performance of the group to detect new areas where performance may be improved.

**Problems and Causes**

At this stage it may be useful to consider the difference between problems and causes; it is essential for the supervisor to be able to separate the two. We will illustrate the distinction between the two by the following case history.

An analytical service was felt by one of its client sections to be consistently failing to provide an efficient service by not meeting deadlines. The reasons were considered to be the unhelpful attitude and lack of organizational skills on the part of the technician operating the machines. The response was to ask a higher graded technician to take responsibility for the service and retrain the original technician. Despite this the complaints continued, for the problems had been identified rather than the causes.

Among the causes of the problem was the appointment of a too low graded technician to perform the service and a degree of retraining was necessary, but the real problem was the submission of more samples than the technician could deal with, and the tendency of the client department to submit a large batch late in the afternoon. The appropriate action would be to implement a more formal booking system with proper request forms and the transfer of an additional member of staff to help at peak periods.

The supervisor needs to decide objectives for the unit not only in terms of performance (as will be discussed later) but in respect of problems, the prime objective being to solve the problem and produce the required results.

## Problem Solving

The line supervisor will continually be faced with the need for problem solving and will need to make decisions. In most cases it will not be practicable to refer the problem to higher authority and await a decision as action is often taken at the shop floor or bench level by subordinate staff when immediate guidance is not available.

While it is important that decisions are not delayed it is essential that they are made on the basis of adequate factual information. Where it is not possible to gather all the necessary information, the supervisor must base the decision making on past experience. If faced with a new problem, however, this method may not be effective. When faced with an equipment or procedural fault the use of a trial and error technique may be acceptable. In the case of a human problem this approach is not possible.

The use of standards is accepted as normal practice in the technical aspects of laboratory work, enabling comparison between samples, tests *etc*. Similar monitoring exists for such items as autoclave runs in biological laboratories, and quality control departments which enable faults to be traced back to particular shifts, staff or procedures. Similar procedures may be acceptable if adopted to suit the supervisory function, although the style of approach must be changed considerably, for while raw products/vaccines may be treated as batches, staff should be treated as individuals.

The basic procedures consist of defining and selecting standards, with agreed methods of monitoring performance. Monitoring should include examination for deviation from the standard or expected result, investigation of the reasons for the difference; looking for the causes not just the problems, investigation of reasons for the differences and adjustment of procedures accordingly. There are additional complications where the deviation is caused by the behaviour of subordinates.

## Group Discussion

The use of group discussion methods may be applicable to those areas where a team is involved. In such discussions ideas and suggestions are made and discussed by all members of the group. This technique is based upon value analysis and a questioning approach, its principal values being to produce unorthodox answers to problems and to bond the group into a team. One of the disadvantages associated with the use of such techniques is the amount of time required to investigate each suggestion. However, tea and coffee breaks may be used by the skilled supervisor as informal situations for group discussions without encroaching upon working time.

The major difficulty in the making and implementation of decisions facing the supervisor may not be the physical aspects of the problem —organizational resources, cost, availability of plant *etc.*—but the effects of the decision upon people. The human relations aspects may be, and often are, overlooked not only by line supervisors but by higher management who often give the impression of making decisions in complete isolation from the working situation.

A balance must be maintained between the physical aspects, the requirements of management and the needs of subordinates. Such a balance can only be achieved if the supervisor makes a conscious effort to consider these aspects of each problem and the consequences of any decision upon each of the three. It is quite possible that a number of solutions will exist to every problem with each solution having different implications in terms of the effect on, and resultant motivation of, staff.

In practice the majority of technicians will have experienced serving under a supervisor who failed to achieve such a balance either through a disregard of the human relations approach to management or as a result of lack of training in supervisory skills. Such supervisors often add to a problem by attempting to impose a solution upon staff or, in complete contrast, by adopting a *laissez-faire* approach which fails to provide any leadership at all.

**THE NEED TO PLAN**

The technical supervisor will be faced with a wide range of problems and varying degrees of pressure to solve them or produce results. Some situations will require immediate actions while others will benefit from deferment or delegation. Although aware of these facts many supervisors will argue that they do recognize the problem, are aware of the need to plan, but are unfortunately too busy dealing with the current situation to undertake detailed planning in respect of future problems. This is a short-sighted view for although planning takes time it should save the supervisor time in the long run. This is particularly so in the case of recurring problems where a planned and developed solution or routine is considerably more efficient than attempting to devise a fresh answer each time the problem arises. In such cases while planning may increase the pressure on the supervisor during the initial stages it will not only provide a saving in time in the longer term but lead to a consistent response and offer the opportunity for delegation.

It is difficult to plan for the unexpected, but within the technical sphere many of what are often considered to be unexpected problems may be anticipated by the supervisor who takes the time to plan.

## HOW TO PLAN

Realistic planning can only occur on the basis of adequate information, without this planning may be more accurately described as guess-work or wishful thinking. The information needs to be collected and collated in a form suitable for the supervisor to base forecasts and set objectives.

### Forecasting

Forecasting may be defined as a logical attempt to predict future events based upon information and experience. The supervisor will need to consider not only the physical aspects of the problem but the human relation aspects touched upon earlier in this chapter and considered in greater detail in Chapter 5.

### Planning

If forecasting is the 'what' stage, planning is concerned with the 'where, when, how and by whom' stages. The supervisor will need to decide upon the objectives for the group and later of the individual members of the group on the basis of the resources available and the demands made upon those resources. Often a number of these factors will in fact be imposed from outside such as fixed lectures, college timetables, field trials *etc.* but hopefully the supervisor will have been consulted when such organizational programmes were devised. Objectives, as well as being essential to effective planning are most useful to staff as they indicate the aims of the organization or group and, perhaps more importantly for the supervisor, allow staff to work with less detailed instructions.

Objectives must be realistic in terms of time, staff capabilities and resources. The supervisor should inform management if the group is set unrealistic targets so as to avoid the damage to morale which results from failure to reach objectives. Where management will not revise the targets the supervisor will need to take action before the short-fall occurs to provide secondary or alternative group objectives.

The objectives set up by the supervisor may be in a number of forms:

• *Problem Solving.* These may be related to technical problems, productivity, quality control, motivation or individual members of staff. The objectives being simply to solve the problem and produce the required results.

• *Performance Objectives.* (i) These may be task objectives requiring the completion of a given task in a certain time while maintaining standards within agreed or acceptable limits, requiring continuous monitoring of results in the defined areas. Performance objectives could also be applied to staff problems *e.g.* turnover or absenteeism.

(ii) Standard increasing objectives are those related simply to increasing performance or developing procedures and chances to improve output. Such objectives should result from a conscious decision on the part of management or the supervisor. Unfortunately the need for the implementation of such objectives is often imposed upon the supervisor as a result of staff cuts.

• *Skill Increasing Objectives.* The supervisor should always be considering ways to increase both supervisory and technical skills within the group. Such objectives may be used to improve the effectiveness of the group and to develop individuals within the group.

• *Process Objectives.* These are objectives designed to meet the broad aims of the organization for the group which are to a considerable extent encompassed in the above. These may include the preparation of test schedules, controlling and processing work, staff motivation and linking the efforts of separate groups.

## KEY RESULT AREAS

It is useful to break the objective down into components or stages and highlight the Key Result Areas (KRA). These are the areas within which good results are essential to the achievement of the objective. Good performance in other areas may be desirable but not essential.

Having decided on what must be done in the key and other areas the supervisor needs to plan how it is to be done, which member of staff will undertake the activities at each stage and when they are to perform the tasks. Deciding on the correct sequence and scheduling of individual tasks within the overall plan may be of prime importance. The supervisor should also calculate the cost of the plan in resources, time and money.

## Control and Feedback

The selection of appropriate control and monitoring procedures are essential for effective planning. Monitoring will need to cover results, quality control and financial aspects (Figure 4.1). The feedback in respect of performance costs and views of staff allows reappraisal of the plan and the inclusion of improvements as necessary.

The use of reports are generally accepted as a necessary part of providing feedback. The supervisor should take care to ensure that all records and reports used are both necessary and seen as necessary by all members of staff involved in their use.

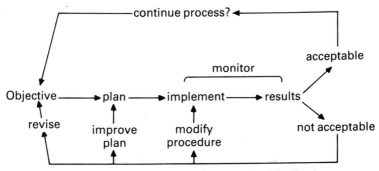

**Figure 4.1** *The planning cycle: control and feedback.*

## Allocation of Duties

Within the group it is necessary to maintain a balance of tasks appropriate to the grade, expertise and career expectations of group members. In allocating work the supervisor should ensure that the mundane or unpleasant jobs are shared as much as possible within these limitations.

The work should also be controlled so as to avoid 'empire building' among subordinates and to prevent subordinate technical staff being incorporated into the 'empires' of academic or scientific staff where separate career and reporting structures are operated. Such empire building and poor work distribution cause ill feeling amongst staff and affect group motivation.

A further difficulty associated with work distribution that the supervisor must avoid is the tendency to place undue emphasis on their own
primary group particularly when allocating the more interesting tasks,
or additional financial reward, *e.g.* overtime or conference work. This
problem is often encountered where the supervisor has been promoted
internally within the establishment.

The supervisor must also ensure that the career development needs
of staff are satisfied. Subordinates must be provided with the chance to
learn new techniques, take responsibilities *etc.*, in preparation for
progression to higher grades.

## PLANNERS AND CHARTS

The majority of supervisors will be familiar with the use of planning
charts for recording staff holidays and histograms or bar charts (if only
from their college course).

In this section we will consider the use of charts systems as an aid to

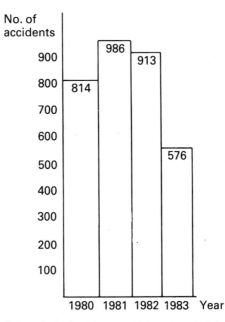

**Figure 4.2** *Simple bar chart. Accidents requiring more than just local first aid in
10 British universities.*
(Data from Kibblewhite, *The Safety Practitioner* **2**, no. 12, December 1984.)

the supervisor in planning work, monitoring progress and controlling staff.

**Histograms**

- *Simple Bar Chart*. These bar charts provide a simple method of representing data to allow comparison of different values. Figure 4.2 shows the use of such a system to illustrate accident rates but could as easily be used to record production data, experimental results or student numbers.
- *Component Bar Chart*. The component bar chart (Figure 4.3) allows comparison of the various components making up a total.
- *Multiple Bar Chart*. These also offer a means of comparison and

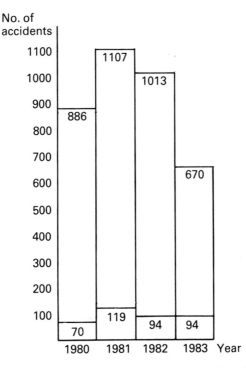

**Figure 4.3** *Component bar chart. Accidents to staff in 10 British universities. The number of notifiable accidents as a component of the total number of injuries requiring more than three days absence from work.*
(Data from Kibblewhite, *The Safety Practitioner* **2**, no. 12, December 1984.)

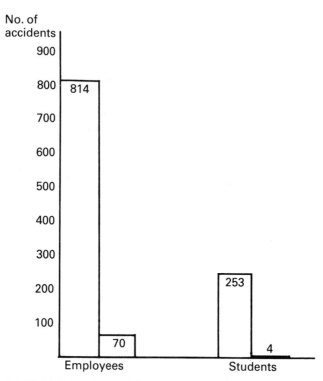

**Figure 4.4** *Multiple bar chart. Multiple bar chart showing accidents to staff and students during a given year.*
(Data from Kibblewhite, *The Safety Practitioner* **2**, no. 12, December 1984.)

may be used for two or more separate items. Figure 4.4 shows accidents for staff and students.

● *Graph.* The conventional graph is best used to show results which form a continuous pattern such as group commitment in relation to involvement (Figure 4.5). Within laboratories they are frequently used to derive values from results or to monitor quality control. One of the limitations of the conventional graph is that of comparison between two sets of results. In some cases this may be overcome by superimposing the two lines on the same graph as in Figure 4.6.

● *Gantt Charts.* Gantt charts take the form of bar charts with horizontal rather than vertical bars. They are frequently used as holiday planners and are available commercially in this form as in Figure 4.7.

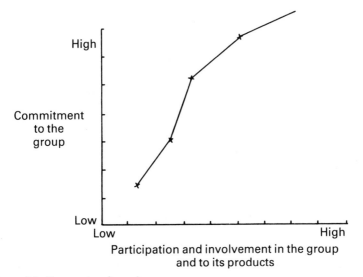

**Figure 4.5** *Conventional graph.*
(After Shtogren. 'Models for Management: The Structures of Competence.'
Teleometrics Int'l.)

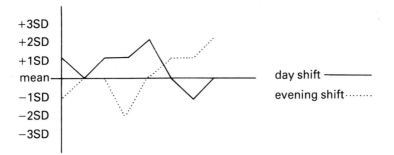

**Figure 4.6** *Conventional Graph allowing comparison of two sets of data as
exemplified by a quality control sheet showing deviation from mean result
between two shifts.*

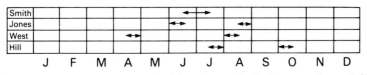

**Figure 4.7** *Gantt Chart used as a holiday planner. Such charts are also useful in
work planning and in monitoring progress.*

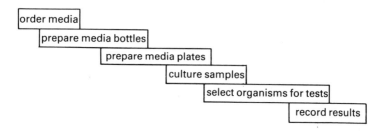

**Figure 4.8** *Sequence bar chart for the preparation of media and identification of samples.*

Such charts are also useful in work planning and in monitoring progress of specific tasks. Figure 4.8 shows how the tasks involved in an operation may be shown on a Gantt chart.

**Special Planning Methods**

There is a profusion of special planning techniques available to management based upon what is variably described as the systems approach, operational research or systems analysis methods. These are particularly concerned with decision making and are based upon mathematical and computer models to improve the decision making process.

The actual preparation and implementation of such systems are the responsibility of specialist staff, but the logical approach to planning adopted by the proponents of these techniques may be used by the line supervisor faced with problems associated with complex tasks.

Two of the most frequently discussed methods of planning applicable to the supervisory level are Critical Path Analysis (CPA) and Programme Evaluation and Review Techniques (PERT). Both of these systems were developed in the 1950's to meet the needs of the US forces to control large projects. In a simplified format they provide the supervisor with a means of visualizing work sequences and controls.

These techniques involve the supervisor in:

● Identifying each separate activity involved in a job along with the time estimated for its completion.
● Arranging these activities into a logical sequence.
● Calculating the resources needed for each element. This enables the cost of proceeding with or abandoning a task to be seen and measured at every stage.

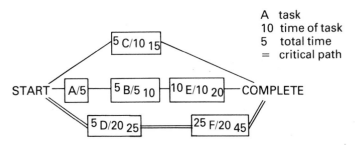

**Figure 4.9** *A network analysis of six tasks in a job. Total time to complete the job 45 units.*

When the separate activities have been identified and ordered they are displayed in the form of a network chart such as the bar charts already discussed or network analysis systems as shown in Figure 4.9. Some of these activities will delay completion of the whole job sequence and these form the critical path.

## MANAGEMENT BY OBJECTIVES

Management, or more correctly in the context of the book, supervision by objectives (SBO) is a technique which can be used most effectively in many laboratories. At its simplest, it entails the use of agreed objectives for the group and each individual, with assessment of performance based upon the ability to achieve the objective (Figure 4.10). This style of indirect rather than direct supervision is favoured by many technicians who, being independent by nature, function

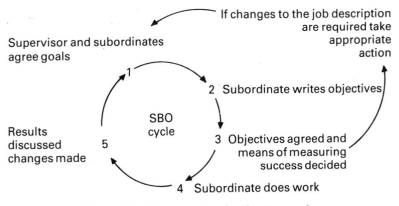

**Figure 4.10** *The supervision by objectives cycle.*

better when not subjected to close control. By its very nature SBO is not appropriate to all technical jobs or to all grades of technician. For the technique to operate well there needs to be effective communication between the supervisor and subordinates and this requires a degree of maturity in both parties. For similar reasons there are many supervisors who find that they cannot adopt the necessary consultative style or accept the considerable flexibility that SBO offers to subordinates.

There may also be a number of situations where it is necessary for the supervisor to know precisely what subordinate staff are doing at any particular moment and in such cases more control will be exercised than is offered by staff working within agreed guidelines.

The technique may lend itself best to the situation found in research laboratories or service sections where the technicians are often not closely supervised in any case and may be more integrated into the team than is the case in other laboratories. Before considering Supervision by Objectives in more detail we would highlight one danger which may not be immediately obvious. If the technical staff of only one or two groups within a building adopt the system of working solely to objectives they may show a tendency to work irregular hours based round the work they are doing. While such flexibility may be welcomed within their own group it may cause ill feeling amongst other staff who witness late morning arrivals but are unaware of the late nights. Where the supervisor has responsibility for all the technical staff in the building this situation might be improved by introducing a form of flexitime for all.

Within the management context MBO may be defined as a means by which the supervisor's functions are integrated into a cycle in which supervisor and subordinates agree targets, work to achieve them, assess progress and make modifications necessary to improve performance (Figure 4.10).

## STEPS IN SUPERVISION BY OBJECTIVES

In order that the supervisor uses the Supervision by Objectives Cycle effectively it is important that a logical approach is followed as suggested below.

### (1) *Agree Goals*
The supervisor discusses with subordinate staff, individually or in small groups, the role of the section in the overall plans of the organization together with the broad objectives required of the unit.

The work necessary for the group to achieve these targets is considered and broad objectives agreed for each subordinate, or team of subordinates in general terms.

Where a job-evaluated grading structure is operated, care must be taken to ensure, at this stage, that the work to be undertaken by individuals will not change their job description in such a way as would affect their grading, unless the supervisor will support and has the authority for the job to be upgraded.

(2) *Set Objectives*
The subordinates then prepare their own detailed objectives in draft form for discussion with the supervisor. At this stage modifications are made to ensure that the objectives meet the needs of both the individual and the group as perceived by the supervisor.

(3) *Set Standards*
Once the objectives are agreed the supervisor and the subordinates decide upon criteria against which success can be judged. This may be on the basis of the number of tests conducted, number of animals bred, degree of accuracy *etc.*

(4) *Perform Tasks*
Subordinates perform the tasks, monitoring their own performance as they do the work to ensure that they are on target.

(5) *Review*
The supervisor and subordinates review the results obtained against the agreed objectives. Agreement is reached on areas in which improvements may be made and whether the original objectives were realistic. New objectives are set, old ones abandoned and the cycle continued.

## A SUPERVISOR INITIATED APPROACH

The traditional view of management by objectives is that it will be instigated by management. However in many laboratories the supervisory technician will find that 'management' will be disinterested, and the technician will thus lack the managerial support implicit in the normal view of MBO.

In the steps given above an allowance has been made for this lack of managerial interest by the proposal of a system of supervision by objectives for use within the group. Hunter in *Supervisory Management* (Reston Publishing Co.) also recognizes this problem and suggests a modification of MBO appropriate for use within a science department which would involve the management in the process. The supervisor prepares draft objectives for the groups under his control as in 1 to 3 above. The technical supervisor then discusses the draft objectives with the manager, scientist or head of department without mentioning MBO or other management terminology. This may be done without actually showing the manager the written objectives, the supervisor merely indicating that he wishes to look ahead and discuss the future. The supervisor can raise the objectives one at a time, emphasizing tasks to be completed rather than objectives as such. During the discussions the supervisor can get agreement on the proposed objectives ensuring that the necessary resources will be made available. It may be possible to agree on measuring methods during these discussions as well as fixing the date of future meetings for the purpose of reviewing progress, but it might initially be advisable for the technician to conduct his own review of results in order to avoid appearing too efficient or keen on management techniques. The results can then be discussed individually with the manager and modifications to the objectives agreed.

## COMMUNICATION

It has almost become a cliché that effective communication is essential to efficient management. Despite, and perhaps because of this, little time is spent in reviewing the effectiveness of the communication structure or in deciding the most appropriate methods. Figure 4.11

**Figure 4.11** *Pyramidal control.*
Modified from 'The Principles and Practice of Management'.
Edited by EFL Brech Longman.

shows a typical reporting pyramid indicating the frequency of reporting at different levels.

The technical supervisor will need to have the ability to communicate both with subordinates and supervisors and, in addition, to specialists, clients or patients *etc.* It is also necessary for the supervisor, as recipient, to understand communications directed towards himself and to appreciate the reasons influencing the choice of communication methods. In each particular situation the supervisor will need to appreciate with whom, and how much to communicate as well as the use of the most appropriate method.

Within the organization there are both formal and informal systems of communication.

**Formal Communications**

These may be either verbal or in writing. Formal verbal communications should normally follow the organizational or reporting structure of the establishment, in order to avoid by-passing and subsequently undermining the authority of supervisors (see Chapter 1). The most common form of verbal communication will be that between supervisor and subordinate, relating to setting and achieving objectives, *i.e.* task orientated. Such verbal communication has the advantage of being two-way, at least in theory, offering the possibility of a meeting of minds and agreement as to what is required. The disadvantage is that there is no permanent record of what has occurred unless one party circulates a memo or note.

The early stages of disciplinary procedures may provide for a formal counselling interview, usually preceded by informal processes, to be given by a supervisor or the Personnel Department and it is usual for such interviews to be noted.

Formal written communication can take a number of forms and have relevance to the organization as a whole, the group or an individual.

Information relevant to the organization as a whole will include rules and procedures relating to employment or safety matters, general advisory information in respect of welfare, conditions, services, agreements with trades union etc. These may be sent to all staff, group supervisors or displayed on notice boards.

Signs and notices are becoming increasingly common in the area of laboratory safety. The number of signs, memos *etc.* should be restricted so as not to swamp the recipients, and they should be enforced. All too often safety signs and memoranda, once issued, are ignored.

At the group level written communications may supplement or reinforce verbal information, provide detailed working procedures, technical protocols or take the form of laboratory reports. Written communication directed to individuals may relate to their terms and conditions of employment, including records of counselling or disciplinary procedures.

## Informal Communication

At the organizational level the 'grape-vine' is the term applied to the informal communication system which tends to be quick if not always accurate. The use of the grape-vine can have a number of disadvantages. It often leaks management intentions, particularly when management fails to communicate effectively with staff, and can undermine the role of supervisors as a link in the information chain. Where the information on the grape-vine is inaccurate or incomplete it can cause concern and ill feeling among staff. While it is possible to deplore such instances they only arise in the absence of an efficient formal communication system.

The supervisor should be aware of the grape-vine and use it to keep informed about the grievances and concerns of staff, to assess their views and to detect rumours which can then be countered if necessary. Under certain circumstances the supervisor can also use the grape-vine to detect and deal with the source of destructive or hurtful gossip, although it may be best to deal with such problems through the informal stages of the official procedure.

Negotiations and discussions between Trades Unions and Management will be of both a formal and informal nature, with emphasis being on the informal where relations are good. The supervisors' most frequent point of contact with trades union representatives within laboratories may be with safety representatives, and supervisors should be clear as to their precise responsibilities in respect of such discussions.

## Communication Problems

It may be argued that most of the problems relating to communication arise from failure to communicate effectively. This may be due to a lack of communication, to the inability of individuals to express themselves adequately, or lack of comprehension on the part of the recipient.

The inability to express the message effectively can either arise from an unfortunate choice of words, the use of jargon or officialese, the absence of sufficient technical knowledge or understanding of the situation 'on the ground'.

Where communication failures occur in using verbal methods, providing the mistake is detected there is an opportunity to rectify the situation. For this reason supervisors are advised to watch for reaction when holding discussions with staff and to sum up at the end on the lines of 'we have agreed on . . .'.

In using written methods the objective should be to keep the sentences short and the message clear. Objectives presented as a list are often easier to follow. Problems may arise within larger organizations where documents may be prepared in terms that apply to as many groups of workers as possible, with the result that they are generalized and not seen to apply to certain specific groups or situations. Such problems could be avoided if the documents were circulated via the supervisor who is given authority to interpret them in a form appropriate to the group, if necessary notifying management of changes made.

Communication does not necessarily improve with experience but it can be taught, and should be included in any programme of supervisor training. One of the major problems is that effective communication is not solely confined to passing on information but in developing the right working relationships where employees are prepared to receive, act upon, and provide feedback on information. As such, communication is essential to the maintenance of the working team but it must be a two-way process. It does entail the supervisor being prepared to listen to staff, to allow them to express their views fully and to accept their views where possible. Once the supervisor or manager decides to impose their own view or to demonstrate dominance and authority, there is little hope of effective communication within the modern employment situation.

Effective communication requires a systematic approach similar to that adopted when planning any other aspect of work and may be considered in three stages. The first of these requires the supervisor to be clear as to the purpose of the communication, whether to obtain improved performance, introduce a new procedure, monitor progress or to pass on safety information.

The second stage consists of research. As with other supervisory procedures the purpose of this data research stage is to ensure that all the relevant information is at hand including, if appropriate, the employment details of individuals. This is particularly important where the supervisor is involved in disciplinary action, as acting incorrectly can have long term consequences both for the individuals and the credibility of the supervisor.

The third stage consists of selecting the most appropriate method of delivery *e.g.* written, verbal, formal or informal. In a number of different situations this will be pre-ordained. To avoid confusion work procedures should always be in writing as should modifications to safety policy. The initial stages of disciplinary procedures should always be verbal and informal, *e.g.* counselling, and will be laid down in appropriate agreements with Trades Unions.

## Reports

Reports usually associated with results obtained from laboratory tests or samples will be one of the most familiar methods of written communication to technical staff. The technical supervisor should encourage the regular use of reports as part of the supervisory function even when these are not specifically required by the organization. Those establishments which do not use 'reporting' systems tend to suffer from poor communication, or at least one way communication as memos *etc.* flow down from management, without adequate provision for a response from staff.

### The Reporting Pyramid

Within many educational laboratories communication problems may result from the lack of an effective reporting procedure between the different levels of the organization. The reverse may be true in private sector laboratories in which attempts to impose tight managerial controls may result in a profusion of reports. Obviously the supervisor will have little control over the reporting systems required by management, but within those areas under the supervisor's direct control, attempts should be made on a regular basis to justify all reports used, to cut out duplicating and overlapping reports, and to simplify those retained in use.

At one time it was felt that the use of microcomputers and word processors in the laboratory would lead to greater use of systems analysis and a consequential reduction in the number of reports. In practice, however, by increasing the ease with which information can be stored, they may be adding to the total.

## TYPES OF REPORTS

### Routine Reports

These may include reports on:

- personnel; *e.g.* as annual progress reports and work, college, and absence reports,
- work progress, quality control reports and analysis of results,
- incident and safety reports on accidents and near misses.

### Special Reports

These relate to non-routine matters perhaps making the case for new equipment, more staff or regrading of a post. These may require more thought and consideration than routine reports as there will be a need to develop an argument and make recommendations. The writer will also need to anticipate arguments against the case being made and answer them in advance.

### Report Writing

A well written report will provide the supervisor with an effective means of communication with those in authority and, as such, should only be submitted after careful thought and preparation.

Report writing may be considered in three stages: preparation, drafting, and review.

- *Preparation* (i) The writer will need to be clear as to the purpose of the report, and it helps if this is written in one short sentence at this stage. Subsequent work can be assessed against this to avoid the inclusion of unnecessary points. Where the 'terms of reference' are provided by those in authority, or a committee, the author should ensure that these are correctly understood.

(ii) Collection of Material. The collection of information may in fact be the most time consuming part of writing the report. The use of index cards for recording notes when collecting information offers an easy means of arranging the pertinent facts in a logical order prior to commencing the actual writing of the report.

- *Drafting* (i) Having arranged the facts in order it is now possible to decide on section headings. These should identify the subject matter and be unambiguous.

(ii) Consideration should be given to the use of diagrams and tables as an effective way of presenting information. These may be included either in the body of the text or as appendices at the end of the report. The advantage of the latter is that the report itself is kept as short as possible and the flow of the text is not disturbed.

(iii) The initial draft of the report can now be prepared, which should consist of:

(a) A brief *introduction* giving the definition of the subject, terms of reference and details of those for whom the report is intended.

(b) The *main body* of the report divided into sections. As an aid to clarifying the sections, or individual paragraphs, these should be numbered.

(c) *Conclusions* and *recommendations*, including if possible the names or positions of those who will be required to implement the recommendations.

(d) A list of *references*, as in a scientific paper any appendices.

(e) A *synopsis*, which should be as brief as possible and may be placed at the beginning of the report as an aid to those who may not have the time to read the full text. This is particularly useful where the report is to be considered by a committee.

## COMMITTEES

Jay, in 'How to run a meeting' (Video Arts Ltd.) identifies three broad categories of meeting. These are:

• *The Assembly*: attended by one hundred people or more where people listen to lectures given by the main speakers.

• *The Council*: forty to one hundred participants, but retaining the use of main speakers while allowing greater involvement from the others attending.

• *The Committee*: preferably five to seven people with a useful maximum of twelve, in which all contribute fully.

The technical supervisor in many fields of employment will find ample opportunity to become involved in committee work within the department, *e.g.* safety, and outside specialist committees *e.g.* professional bodies or trades union. The 'Istox Laboratory Management Survey' (1979) showed that technicians were involved in inter-organizational committees in 62% of laboratories, external committees in 44% and trades union committees in 30%. The actively involved supervisor will certainly be on several different committees, both to advance his own position/section or to gain knowledge. It is often said, by those who seem to collect committees that they do not like committee work and would rather be more involved at the bench. In practice this is not meant to be taken seriously as committees are, both in principle and practice, an essential part of management and for a supervisor to adopt this view would be an abrogation of managerial duties. The committee provides a means by which a number of people can bring together their expertise and reach a balanced or representative decision. On the other hand committees can be used as an excuse to avoid or delay decisions, to talk and not act or to spread the blame for unpopular decisions. The consensus style of management, popular in some sectors during the seventies, based upon a management committee or team, often failed to provide a clear lead, fudged responsibility and led to ineffectual compromises as well as being expensive in man hours.

The effectiveness of committees depends upon them having clearly defined terms of reference, an effective chairperson and informed members.

**Types of Committee**

Committees may have executive authority, be of an advisory nature, or be consultative. They may masquerade as working parties, advisory groups, steering groups, *ad hoc* groups or sub-committees, depending upon their size, function and the politics involved.

● *Executive Committee* in which authority is vested to make decisions and where the committee is collectively responsible for them, *e.g.* the board, departmental committees in science departments or trades union branch committees.

● *Advisory Committees* submit information and advice in the form of a recommendation for a manager, specialist or another committee to act upon. In many organizations the safety committee will act in an advisory capacity being responsible to the Safety Officer, while in others the position will be reversed with the Safety Officer (or Advisor) being responsible to an executive Safety Committee. The distinction between the two roles is of considerable importance to the laboratory supervisor, both from the committee aspects and, more importantly, the day to day role of the Safety Officer, where the distinction between advisory and executive authority can determine 'where the buck stops' if things go wrong.

• *The Consultative Committee* often provides the means by which management discusses matters with their employees' representatives (Trades Union or Staff Association). Information may be exchanged by both sides and, in theory, this is meant to provide an effective means of communication between employer and employee. In practice, it may degenerate to the level where little communication is involved, with both sides stating their own position and remaining at the table while the other side states theirs. Such committees do not usually vote although on non-contentious topics they may reach a joint decision.

### Committee Meetings

The nature and effectiveness of a committee is influenced by a large number of factors, the principal four being:

• *Composition.* Committees consisting of people who work together *e.g.* departmental or section committees, tend to differ in effectiveness from those that are made up of people who only come together for the committee meetings. Where the latter occurs, more power tends to rest with the officers and secretariat than in the former case.

• *Frequency of Meetings.* In general, the greater the frequency of meetings the more effective the committee will be, provided this is not taken to the extreme where meetings replace work. Where a committee meets infrequently, perhaps once or twice a year, most action will tend to be taken on the initiative of the officers and reported, after the event, to the committee members.

• *Motivation.* The motivation of the committee members will have a considerable influence on the working of the committee, and the atmosphere at a meeting. Members may share a common objective, have widely differing goals, or be competing with each other for resources *e.g.* money.

• *Decision Process.* This was touched upon earlier where the three types of committee were considered. Decisions may be reached by consensus, vote or by the decision of the chairperson alone.

New committee members should be provided with a copy of the constitution or terms of reference before attending their first meeting, together with the names and addresses of other committee members, the minutes of the last meeting, and the agenda for the next meeting. These should be studied and any members known to the new member personally contacted prior to the meeting. It is often advisable for the new member not to speak too much during the first meeting, but to listen to others and comment on their points. It is also useful to take notes and compare them with the minutes that will eventually be produced. Inaccurate minutes, either as a result of incorrect recording or omission, should be challenged, as once accepted by the committee they provide the official record of the meeting.

CHAPTER 5

# Motivation

## THEORY

Although this book is concerned with the practice of supervisory skills rather than management theory, a limited knowledge of relevant theory is of value to the supervisor in fulfilling their role and in understanding the attitudes of management. A brief summary of some of the main theories is given below.

The nineteenth century view of workers was based upon the assumption that they worked solely for money and were both stupid and lazy, capable of working only under strict supervision.

### Taylor

The principal exponent of this view in more recent times was F. W. Taylor ('Scientific Management', New York; 1911) and his disciples may still be found alive and well, working in many laboratories today.

Taylor introduced the term 'scientific management' and gave four principles concerned with the 'science of work'. These were:

(i) observation and measurement to establish a fair day's work,

(ii) scientific selection and training for the work,

(iii) the co-ordination of the trained workers with standardized jobs to ensure low production costs for the employer and high wages for the work based on Work Study Methods,

(iv) the need of employees to cooperate with employers in the investigation and introduction of new methods and procedures.

Under this system the supervisor would be involved in functional management, being responsible for a specialist function or activity *e.g.* maintenance, production. Such a system could result in a worker having a number of supervisors each responsible for a specific function.

By a small stretch of the imagination this structure may be applied in a modified form to some educational laboratories, where technicians maybe responsible to a senior technician for bench work, a departmental superintendent for holidays and time keeping, an academic or scientist for a research programme, and possibly another for practical classes.

## Mayo

Studies carried out at the Hawthorne (Chicago) Works of GEC in the 1920s–30s demonstrated that work is a group activity and that people work better when subjected to interest and concern. In the case of those in the initial studies, the interest was shown by experimentalists but normally it would be by managers.

The Hawthorne experiments consisted of introducing breaks as part of the working pattern which increased productivity; however, when the breaks were withdrawn the rate of production increased further. An interview programme was instigated to ascertain whether workers' attitudes and morale increased as the staff responded to management's interest in them. The final stage consisted of work on group behaviour from which it was found that group solidarity took preference over financial reward.

Mayo's work, although possibly starting the human relations approach to management, did not alter the basic principle that managers should manage and workers work. Mayo proposed that management used new techniques to improve the performance of the workers, stressing the importance of social factors and the role of the group in determining output.

## Argyris

Argyris developed the theory that conflict arose between workers and the organization as they had different goals or objectives, with the organization demanding that workers subjugate their own needs to comply with the structure of job requirements. He claimed that all workers have a potential which, given the right environment and support, can be developed for the benefit of the individual, the group and the organization.

**Maslow and McGregor**

Maslow and McGregor demonstrated that people were a complex mixture of motives and reactions. They recognized that workers were capable of decision making in their own areas, and that such decisions were likely to be better informed than those made by managers at some distance from the problem.

Maslow, in his work on peoples' needs, arranged them in a hierarchy ranging from the basic physical needs for survival (food) to the needs for self-fulfilment (Figure 5.1). He argued that when the physiological needs are satisfied, the needs for safety become important. Once the needs for safety and security have been secured, the need for social acceptance by the group assumes priority. This in turn is replaced by the need for self-respect and self-fulfilment or development.

Physiological → safety → social → self esteem → self fulfilment

**Figure 5.1** *Hierarchy of needs.*

Although Maslow's work has been criticized as not being directly relevant to the employment situation, studies by other workers suggest that it may in part apply. Evidence suggests that workers at the lower levels are most concerned with pay and security while those at the more senior level, and consequently with better pay, are more concerned about achievement, success and self-actualization. Argyle in 'The Social Psychology of Work' (Pelican Books 1972) suggests that this could be due to more ambitious men reaching higher positions within the organization.

**Muray**

Muray also stressed the importance of people's needs and classified them in terms of their dominant needs *e.g.* power, autonomy, achievement. This work was applied to the work situation by McLelland with particular emphasis on the constructive need to achieve. He showed that those with a high achievement motivation were stimulated to extra effort not only by success but also by failure.

**Herzberg**

Herzberg in the 1950s–60s recognized the importance of motivators, such as recognition, responsibility *etc.* and applied them to motives within the organization. He also felt that other factors relating to working conditions, described as maintenance or hygiene factors, although not motivators in themselves, enabled the motivators to work.

These hygiene factors not only exhibit a positive effect, but may have a serious negative effect upon the motivation of staff who are adversely affected when they are withdrawn.

The working conditions within the organization, or the employment environment, may be considered to include all those work factors other than the job itself. Some of these factors are within the direct influence of the supervisor *e.g.* style of supervision, interpretation of company rules, enforcement of safety procedures, and training. These and other motivators are illustrated in the 'Motivation Swingometer' shown in Figure 5.2.

Other factors are outside the direct control of the supervisor, *e.g.* formal conditions of employment, welfare, pensions, job security, and prospects. Intergroup relationships, already discussed under organizational structure, can also have a considerable demotivating effect.

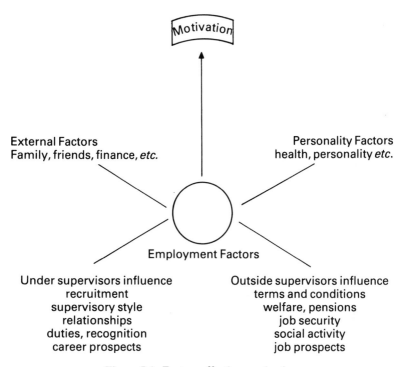

**Figure 5.2** *Factors affecting motivation.*

The degree of common interest and cooperation not only between groups of technical staff but also between the technicians and other job groups can all affect the motivation of the individual within the unit. Positive effects result from co-operation, negative from conflict. Being under criticism or interference from outside can unite the group, and where the interference cannot be stopped the best the supervisor can do is to attempt to limit the damage by creating a team spirit within the section against those outside the group, even at the expense of deflecting it from the corporate plan.

All staff need to be reassured that their work is important, or at least useful, and the supervisor will need to ensure that priorities imposed from outside do not detract from this feeling. For instance where a conference, or other major meeting, is held in the department the supervisor is likely to concentrate maximum effort to ensure that staff are available to provide supporting services. Where this causes disruption to the working of the department, the supervisor will need to ensure that those inconvenienced do not feel that their own normal work is considered to be of little or no importance. Such changes not only affect those directly concerned but other members of the department for whom they work, particularly if service sections are involved *e.g.* photography, stores, *etc.*

### McGregor's X and Y Theory

The attitude of supervisors and managers towards staff is of vital importance and work by McGregor suggests that managers have in the past been wrong in the way they have viewed and treated their workers. He recommended that changes should be made in both organizational structures and attitudes.

McGregor developed X and Y theories in his analysis of people at work.

- *The X Theory:* this dealt with the traditional attitude of employers towards their employees. The view was that the workers disliked and avoided work and responsibility, needed control, coercion, and direct supervision. McGregor argued that the Taylor approach resulted in a self-fulfilling prophecy with repressed workers rebelling against the autocratic attitude of management. This reaction confirmed management's view as to the need for firm control which in turn repressed the workers further.

- *The Y Theory:* This offers the alternative explanation of behaviour, stating that work is natural, that people will work towards organizational objectives by self discipline, rewards and responsibility. The implications of the Y Theory are of major importance to supervisors and the distribution of responsibility within the organization. This may have given rise to the participative style of management.

If the X Theory is adopted, problems of performance and output can be blamed on an intransigent workforce. The Y Theory suggests that failures are the fault of management for failing to effectively organize or motivate the staff.

The suggestion that managers are paid to manage and that failures to 'produce the goods' indicates a failure of management appears more obvious than radical but it still is not fully accepted, at least within the public sector. All too often, while credit rapidly flows up within the system, responsibility and blame when things go wrong seem invariably to work down.

## JOB DESIGN

The actual job in which staff are employed can be a major feature in maintaining the motivation of the individual. The work should be interesting, varied, produce a sense of achievement and be closely associated with a product or service which is seen to be of value to the organization. Jobs should be designed to ensure that the individual's capacities should be fully stretched, but not overextended if stress related illness is to be avoided. Problems associated with overwork are easily appreciated but those caused by lack of fulfilment may become increasingly important if the trends towards deskilling and 'machine minding' discussed in Chapter 14 continue to spread through the technical sphere.

Unfortunately, few technical posts are actually designed to fulfil job design criteria but at least some of them can be found in the work undertaken by research technicians. In service, quality control, or production posts there may be difficulties in meeting any of these criteria due to the routine repetitive nature of the work. Variety may be introduced by means of job rotation but often the supervisor does not have the authority to transfer staff between posts particularly where there is a rigid job evaluated structure.

## Job Rotation

Job rotation is often used during the training period of junior staff allowing them to gain experience in a number of jobs and disciplines. Once the technician has completed the training period and obtained a qualification the rotation normally ceases with the technician being assimilated into a post in a specific laboratory.

Job rotation can be a useful tool if used to give staff a change from repetitive routine tasks, as it adds to their interests and consequently performance. On the debit side, rotation can lead to problems if tasks are left unfinished and the work area is left untidy at the change over, or if individual members of staff do not know what each other are meant to be doing in a poorly supervised section. The need for the supervisor to consider the implications of such problems will increase if Job Sharing Schemes and the use of non-employed trainees, such as the Training Agency (previously MSC) Youth Training Schemes (YTS) gain wide usage. With organization and planning the laboratory supervisor can overcome such difficulties, as similar situations have long been encountered by technicians involved in shift working.

## Job Enlargement

Used carefully, job enlargement can be used to produce a considerable increase in job satisfaction, increasing the work by the provision of more, and preferably different, tasks. Obviously this approach can also lead to problems if not handled carefully by the supervisor. The tasks need to be allocated to the subordinate on the basis of mutual agreement and a desire on the part of the subordinate to increase the workload.

Where extra work is allocated to staff without consultation, agreement or reward, perhaps as a result of redistribution of workloads following staff cuts or redundancies, the effect will be demotivation rather than motivation. An additional problem associated with poorly planned job enlargement in the inability of the subordinate to fulfil the original duties as a result of attempting to complete the additional tasks.

## Job Enrichment

Where subordinates are sufficiently mature, the job can be enriched by increasing the planning and control aspects. Such job enrichment not only offers a most effective means of motivating technicians but results in the supervisor obtaining more time for other supervisory duties, besides providing for the career development of the individual. In job evaluated grading systems, job enrichment is not always possible without affecting the grading of staff.

## Job Design

Job design can also be used to solve other problems that will affect the motivation of staff where difficulties arise from people being unsuited or a poor fit to their job. This situation can arise in two ways. In the first, the individual may be 'bigger than the job' or is in danger of outgrowing the job without the possibility of promotion. In such situations the job content may be increased by job enlargement or by providing additional responsibilities.

Where a job evaluated grading system is operated and management will not sanction promotion the supervisor will need to ensure that the additional work does not take the post into a higher grade within the structure. This may be achieved by ensuring that the extra duties are at the same or lower level than those already undertaken or by selecting duties that are outside of the grading structure. Such duties could include for example, departmental safety officer, committee work, social club duties or even trades union representative.

The second situation is encountered where the job is too much for the occupant either as a result of personality, individual capacity of the worker, or from external factors.

The personality and capacity of the individual depends upon a number of factors including basic drive and the conditioned response to stimuli throughout life. The supervisor at this level cannot be expected to psychoanalyse staff but they should be aware of these basic influences and attempt to review the capacity of individuals so as to design jobs to fit the staff in employment. In cases where the problem arises from the lack of ability of the individual, retraining can be offered to improve performance or alternatively the job can be decreased in size or complexity to bring it within the ability of the individual.

## External Factors and Motivation

The external factors affecting the motivation and performance of staff are innumerable and include family, friends, financial problems, housing *etc.* These problems are frequently only noticed by the supervisor after they have had a disruptive effect as friends and colleagues may be reluctant to report their peers to those in authority. The supervisor should be aware of the influences of the external non-work environment and attempt to create the type of climate at work where they can be approached informally at times of trouble. It is important that the supervisor considers the possible causes of problems if the subordinate's attitude to work declines.

Often counselling and sympathy on the part of the supervisor can help while on other occasions it might prove necessary for duties to be reassigned on a temporary basis. The supervisor should be aware of the advisory and counselling services available within the organiz-ation, or if none are available the facilities offered in the locality, so that it is possible to guide those in difficulties to enable them to find appropriate help. The supervisor must, in addition to providing help and sympathy for the individual, remain aware of the effect of the disruption upon other members of the group. If it becomes necessary to reassign duties the person receiving the additional work must be rewarded, although not necessarily financially.

Other factors within the overall organizational system affect the motivation of individuals. These will include the future plans of the organization, both unfavourable known plans and uncertainty over the unknown. Uncertainty and rumour tend to have a negative effect and the supervisor should ensure that management are aware of the feelings among the workers. The latter is particularly important where staff are in the final year of fixed term contracts as doubt as to the prospects of renewal can have a marked effect upon motivation. Other factors relating to staffing and change can have both positive and negative effects upon motivation.

If the theory of needs is correct, when a need arises people act to satisfy that need. The supervisor will in most cases be unable to identify the needs of subordinates and can only infer the existence of a particular need from the conduct of staff. Indeed people may not even be aware of their own needs or if they are they may not feel able to discuss them with their supervisor. The situation is further complicated by the fact that even if the needs of staff do not change the means of satisfying them will vary depending upon the people involved and the circumstances. Despite these difficulties the supervisor must attempt to satisfy the needs of subordinates as frustrated needs may lead to behaviour directly affecting the performance of individuals (Figure 5.3).

**Figure 5.3** *Job satisfaction.*

## The Human Relations Approach

The recognition of the importance of motivating staff leads naturally to the human relations approach to supervision. This will require a considerable change in the attitudes of the technical supervisor who during his career at the bench will have placed emphasis on technical skills, the application of precise procedures and the importance of exact results. It is inappropriate to attempt to apply such attitudes to the complex relationships that exist between people. Supervisors who are unable to make the necessary change in attitude will encounter difficulties in their relationships with staff that can only have a detrimental effect upon productivity.

If there is any doubt as to the importance of people in the technical or laboratory environment the supervisor need only consider the problems encountered in a normal week. Are these related to techniques or people? Are the causes of trouble or crisis related to equipment or staff? In the majority of cases it is the relationships between people at work that cause difficulties and these difficulties tend to recur.

When discussing motivation we mentioned people's needs. These will now be considered in more detail with emphasis on the practical applications for the supervisor. It is suggested that man's basic needs may produce certain behaviour traits and, in the Process Theory, that choices are made on the basis of past knowledge with alternatives being selected to give the most favourable or pleasurable outcome; the three major determinatives of action being Reinforcement, Need and Incentive.

## NEEDS

### Self Assertion

The need to feel important may be as an individual or as a member of a group. The good supervisor can use this need constructively by providing personal attention to all members of staff, fair treatment, and good training. This requires a flexible approach, judging each situation on its circumstances and not being bound by, or afraid of, creating precedents. A good system of promotion and recognition by means of increased status or financial reward will also help to meet this need.

### Acquisitiveness

This need may be met by the provision of status or possessions, *e.g.* own desk, office area, job title or by the provision of protection, security at work, pensions *etc.*

### Aggressiveness

The natural aggressiveness of people may be harnessed to the good of the organization *e.g.* aggressive and effective sales staff, the section leader aggressive in the defence of the group. It can however be destructive in subordinate staff if it produces a trouble-maker always airing grievances, particularly if these are founded upon a grain of truth and taken up by the union. The skilled supervisor will identify the causes of friction and remove them rather than brand an individual within the group as a trouble-maker and thus create a leader of discontent.

### The Need to Belong

This can be used to positive effect if the supervisor can direct it into the need to belong to, and be accepted as, part of the working group. Individuals rejected by, or working outside of the official group, may create their own grouping with possible conflict of objectives developing.

**The Creative Urge**

The supervisor should encourage staff to become fully involved in the work of the group, to suggest improvements in working methods, to take responsibility for their own job and adopt a flexible approach to the tasks in hand. The supervisor should foster this urge by publicly recognizing the efforts of subordinates. The basic drives, once they become sufficiently strong, will either generate action or be repressed, building up and possibly causing retaliation or other action at a later date. This may result in a disproportionate reaction to a minor triggering item. The supervisor will need considerable skill to recognize the existence of repressed needs and to provide opportunities for them to be redirected and worked-off constructively.

Where poor human relations exist within the organization there may be a number of possible causes ranging from the style of leadership, failure of communication or organizational failures. The latter could include problems associated with organizational structures or poor selection and training of staff and supervisors.

In the section on organizational structures it was suggested that the structure was determined by the objectives, policies and function of the organization, with staff being employed to fill that structure. In practice, the principle of structure first and people second is often ignored to fit the wishes of senior staff who have the authority to indulge in 'empire building'. For example, they may wish to appoint a recent graduate or post graduate to a technician post while they wait for a more appropriate opening, or to find a 'hole' for a long-serving but ineffective member of staff. Such abuses of the structure, while they may be seen as the right of management, cause concern among staff and encourage many to take a cynical attitude towards those in authority.

The organization which is dedicated to a human relations approach to management will endeavour to create and develop competence among staff, not only by developing an appropriate structure but by training, counselling and planned promotion programmes.

Within the group, the supervisor can work to improve human relations in a number of ways which are based more on common sense than motivational theory. These include:

• taking an interest in subordinates, listening to them (not only when they have problems) and praising their efforts;

- not arguing with subordinates. Supervisors can rarely win in such cases for while they may score points or intimidate staff it will be at the cost of demotivation;

- being prepared to admit mistakes while taking care to be constructive and not destructive when subordinates do things wrong, resisting criticising staff directly and attempting to avoid misunderstandings;

- explaining policy and objectives to the group and being seen to support them in conflicts with management or other groups:

- never forgetting the importance of people to the organization.

Barclay ('Technical Managers Activity and Training Survey 1982', Technical Management Unit, Huddersfield Polytechnic) found that people management ranked highest in an overall job activity analysis of supervisors at 92.8%, with interpersonal skills third at 83.8%. He concluded that people and money related problems cause technicians and technical managers most trouble and remain troublesome even after the supervisor has obtained considerable experience.


## JOB SATISFACTION

Job satisfaction is closely linked with motivation and job design. Some factors affecting job satisfaction are shown in Figure 5.3. Experience has shown that the majority of technicians find considerable satisfaction in their work, although in recent years it seems that they are becoming more cynical and critical of the short comings in organizational and supervisory aspects than has been the case in the past.

A survey of 2460 people in America (Gurin, Veroff and Field, 1960) showed that 42% of professional and technical staff were very satisfied with their jobs while 42% were satisfied. This compared with 22% and 39% for clerical staff.

Various methods have been used in attempts to identify those factors that cause job satisfaction, one of which is to ask people to rank different factors in order of importance.

Rankings produced by Jurgensen (1948) gave the following order:

(1)  Security
(2)  Promotion prospects
(3)  Interesting work
(4)  Company

(5) Pay
(6) Co-workers
(7) Supervisors
(8) Hours of work
(9) Working conditions.

Difficulties for the supervisor may arise when staff faced with problems in one area of their relationship with the organization may complain, or vent their frustrations, in other areas. For instance, lack of promotion prospects may result in staff taking an unhelpful attitude to hours of work or working conditions. The factors relating to motivation should not be considered in isolation but taken as part of the whole. It is interesting to note that the supervisor came seventh in the listing above. In other surveys it has varied from third to seventh. As a factor, supervision is thought more important with white-collar workers and draughtsmen than with manual workers. The latter are often pleased to be left alone by the supervisor whereas white-collar staff place importance on interpersonal behaviour.

**Security**

Within the technical sphere, lack of security may come to assume greater importance because of the effects of the recession in the private sector and expenditure cuts in the public. One aspect of this problem, which may cause supervisors difficulties, is the tendency within universities and medical research laboratories towards the increased use of short term (2 or 3 year) contracts for technicians. While the supervisor may not be able to influence the job security within the organization, they should be seen to be using all their efforts to secure the jobs within the group and in helping those whose contracts are not being renewed to find fresh employment either within the organization or outside.

**Promotion Prospects**

Herzberg founded that promotion prospects were one of the more important causes of positive satisfaction but if people anticipating promotion did not get it they showed considerable discontent. Increased pay may be expected to be linked with promotion prospects, as in most cases promotion will result in an increase in salary, but to obtain maximum benefit from promotion it should include increased

status. In those organizations where there is little difference in status
between different technical grades, as in some universities, the two are
seen by the staff as being synonymous. Relative pay is considered to be
closely linked with job satisfaction particularly in those areas such as
technical employment in the public sector where comparisons, both
within and without the organization, are easily made.

Comparisons within the organization are thus of importance to the
supervisor if a high level of job satisfaction is to be maintained within
the group. While the personnel section may place great store upon
job-evaluation as a means of ensuring comparability between technicians in different groups, a problem often arises among technicians
in that such exercises compare job descriptions while technicians
compare the actual work done and their personal knowledge of the
performance of individuals. Discrepancies between the job description and the job may result in considerable dissatisfaction when
technicians see colleagues being upgraded beyond their perception of
the correct grade.

**Interesting Work**

As discussed under 'Job Design' above, there are a number of factors
which combine to make a job interesting. These include variety, which
may be introduced by job enlargement or rotation, the opportunity to
use skills and expertise, a degree of control, and the chance for
interaction with others.

When introducing variety into a job, the supervisor should ensure
that the additional tasks should be complete in themselves, or form
part of a whole which will involve the worker. The addition of a large
number of 'bits and pieces' of work or duties that are beyond the
competence of the individual, are more likely to lead to dissatisfaction
rather than satisfaction.

Self-expression and concern with self-fulfilment are high in
Maslow's hierarchy of needs and from experience we would support
the view that those technicians who are able to make full use of
their skills and expertise are likely to be more satisfied than those who
have less challenging jobs. The amount and pressure of work also
affect satisfaction; too little or too much on a regular basis are to be
avoided.

The degree of control that staff have over their work is also
important. While most technicians will be required to follow standardized methods they may be involved in the selection of the methods, be

given a degree of autonomy in their use, and freedom in the organization of associated tasks.

## The Company

The reputation of the company and its attitude towards both employees and clients can affect job satisfaction. An organization which is concerned with people and is seen to be fair may appeal more to staff than a non-caring organization. Other related factors which affect job satisfaction are the company size, the extent of participation in management, the skills of managers and the personnel policies of the organization.

While laboratory technicians may be employed in large organizations the individual departments or establishments themselves tend to be of small or medium size which, with reasonable personnel policies, should encourage a degree of job satisfaction. The management and supervisory skills demonstrated within the organization on the other hand tend to be limited as the managers tend to be scientists or technicians who have had little or no training for their management role.

## Co-workers

Colleagues and group membership are important for job satisfaction and those working in groups tend to show a greater degree of satisfaction than those working alone.

While the advocates of the human relations approach considered that the group was the main source of satisfaction, surveys by others place it from third to eighth in order of importance. It would seem that more satisfaction is obtained from the group where people work as a team.

The group size is important with smaller groups having a higher degree of satisfaction, possibly as they offer better opportunities to contribute and become involved in the activities of the group.

## Hours of Work

In manual workers, absenteeism increases with the length of the working week but among other groups, such as managers, there is a tendency to work long hours and to take work home while enjoying a

high level of job satisfaction. This may reflect the degree of involvement in, and commitment to, the job. Often within laboratories the
nature of the work tends to encourage a flexible approach to hours of
work and time keeping. Attempts by administrators to impose a more
rigid regime can often have a detrimental effect.

Shift work tends to be relatively unpopular but the degree of
unpopularity seems to be related to the extent of disruption caused to
other activities. A similar attitude would apply to contractual overtime
as is worked in some animal houses at weekends. The extent of
financial and other rewards for such work will also affect the technicians' feelings towards it.

## Working Conditions

Providing staff are fully aware of the working conditions on appointment they will have little effect on motivation unless they change.
Worsening conditions, either real or perceived, can have a negative
effect upon staff. Within the laboratory such changes could include
work in refrigerated areas, incubators, excessive noise or fumes.
As with salaries, dissatisfaction may result from comparisons with
other groups within the organization or as a result of management's
failure to implement changes required following inspections by safety
representatives.

# Recruitment and Selection

In many laboratories there is little advance planning in respect of staff recruitment, with vacancies being filled as they arise, by promotion of existing staff or expansion of the work force. The good supervisor should attempt to meet future recruitment needs by planning the careers of existing staff, and anticipating future product or work-load developments in relation to the objectives of the organization.

Where it is expected to recruit direct from full time education, be it from school, college or university, the recruitment effort should be co-ordinated so as to attract candidates at the time that they are seeking employment, prior to or immediately after examinations, but with appointment being made when the examination results are known. In such circumstances it would be advantageous for the supervisor to establish links with local schools and colleges to ensure that they are aware of the laboratory's requirements and to assist the recruitment process.

## RECRUITMENT

Recruitment may be considered in two main stages:

- deciding on the post to be filled and the type of candidate required to fill the post,
- attracting suitable candidates.

Figure 6.1 illustrates the various elements involved in these stages together with some of the tasks resulting from decisions made during the recruitment process.

## JOB DEFINITION

It is useful to use the opportunity provided by a job becoming vacant to examine the post and job content critically. Such an examination may

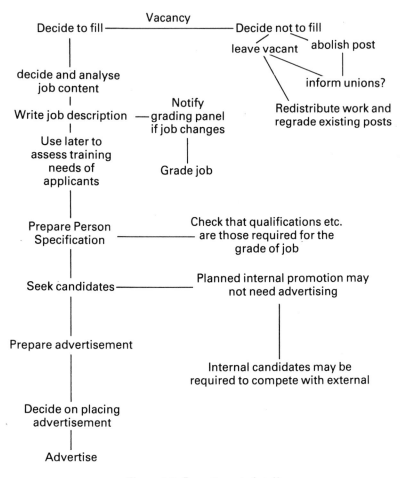

**Figure 6.1** *Recruitment of staff.*

show that the employee due to leave does not actually need replacing, that the work could be redistributed as part of a job enlargement programme or that the post may be 'given-up' as part of imposed staffing cuts in exchange for the retention of a more valuable job in the future.

The existence of a vacancy may also offer the opportunity to modify the job content, to change the grading of the post or to include new equipment/technology in the job description.

## Job Content and Job Analysis

Having decided to fill a particular post the next stage in the recruitment process is that of job analysis including the preparation of a job description and specification as discussed in Chapter 9. In addition there is a need to prepare a Person or Employee Specification to detail the type of person required to fill the vacancy both in respect of job content and the ability to work with existing staff.

## PERSON SPECIFICATION

Job descriptions and specifications are widely used and readily accepted as part of Job Evaluated Grading Systems. Formal person specifications are less widely used but nevertheless may be considered as a valuable part of a systematic approach to recruitment. Perhaps one of the main reasons why such specifications have failed to obtain wider use relates to the intergroup relationship between the personnel section and technical supervisors. Where this relationship is poor there may be a tendency for the specialists in the personnel section to keep technician involvement in recruitment to the absolute minimum; while generalized job descriptions may be kept on a central file the preparation of a person specification would require the active involvement of the technical supervisor. Alternatively where control of the recruitment rests with the technical supervisor and/or scientist they may not be aware of the value of a systematic approach to non-scientific matters. The four main factors in a person specification are shown in Figure 6.2.

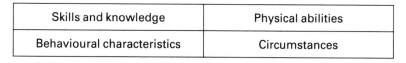

| Skills and knowledge | Physical abilities |
| --- | --- |
| Behavioural characteristics | Circumstances |

**Figure 6.2** *Factors in a person specification.*

## Skills and Knowledge

These are the areas normally covered by qualifications and experience, and at first glance present few problems for the supervisor. However, experience must not be confused with skill. It is quite possible for an applicant to have ten years experience of doing a job badly or of being grossly under-employed.

Skills describe the ability to perform tasks (actually do something) and these tasks may be selected from the job description and broken down into their skill elements. Communication skills, written or verbal should not be forgotten because they are not implicit in the job description.

'Knowledge' refers to knowing about the subject. This is slightly different from 'qualifications' which relate to the ability to pass examinations. There are further problems in respect of qualifications, particularly BTEC examinations, because there may be variation between the programme contents offered by different colleges, both in terms of the actual subjects studied and the contents of syllabuses. To avoid this particular difficulty the supervisor could specify the actual BTEC units that are considered essential in each particular situation in addition to the overall title of the award.

## Physical Requirements

In many technical posts there may be few specific physical requirements, apart from stamina and ability to work under stress. In others there may be a requirement to lift or carry (workshop materials, animal diet) to tolerate heat or cold (working in an incubator or cold room), while in others the absence of colour blindness may be important (electronics, microscopy). In work involving animals, staff need to be free from Laboratory Animal Allergy although this may still develop during their working life.

Age and height limits may also be recorded under this section of the specification.

## Behavioural Characteristics and Character

These are probably the most important factors but are the most difficult to define and assess. Attributes such as aggressiveness, reliability and ability to get on well with others may be desirable.

## Interests

The person's interests may give an indication of the behavioural characteristics where they relate to sports *etc.* but as a heading on the person specification, interests should be considered in the context of the competence of the person to perform the tasks. The supervisor may include membership of professional bodies and associations.

Preferences in aspects of the job or career development should also be considered here as they will need to coincide with those areas of the job that the supervisor sees as developing in the future.

**Circumstances**

These relate to the personal circumstances of the job-holder and are important where the job may impose strains on the individual's home-life *e.g.* irregular hours, contractual overtime, shift work, travelling *etc.*

The final point to remember is that it is not the 'best' candidate who is required for every job but the most suitable or competent. The best candidate in terms of qualifications, intelligence *etc.* may not be as suited to the post as someone who is merely 'good'.

An example of a Person Specification is shown in Figure 6.3, indicating factors under the headings considered above, appropriate to an animal house technician.

### Person Specification—Animal House Technician

**Skills**
(1) Ability to handle laboratory animals without appearing nervous.
(2) Experience in sexing rats at weaning.
(3) Experience in use of automatic cage washers.

**Knowledge**
(1) Knowledge of routine animal house procedures to National Certificate level in Animal Technology.
(2) Knowledge of rules and regulations governing the use of animals.
(3) Limited knowledge of stock control and ordering procedures.

**Physical Characteristics**
(1) Ability to lift cages and sacks of diet.
(2) Freedom from laboratory animal allergy.

**Behavioural Characteristics**
(1) Fond of animals.
(2) Reliable, able to work unsupervised.
(3) Prepared to work weekends.
(4) Able to work as part of a team and communicate with non-team members.
(5) Demonstrates additional interest in the profession e.g. Institute Membership.

**Circumstances**
(1) Lives within easy travelling distance.
(2) Own transport for Bank Holiday working.

**Figure 6.3** *Simplified person specification for an animal technician post.*

## ATTRACTING THE CANDIDATE

### Advertising the Vacancy

The purpose of an advertisement is to attract applications from a reasonable number of suitable candidates. A badly worded advertisement will either attract too many unsuitable applicants or insufficient to provide an effective choice. There are a number of sources of candidates. Internal applications are very important in respect of the motivation of staff and if they form part of a planned promotion policy may result in an appointment without advertising. Alternatively such appointments may be on the basis of internal advertisements on noticeboards, circulars to heads of sections or in a 'house-magazine'.

In addition to contributing to the morale of staff, internal appointments have a number of benefits to the organization. The principal benefit of this being the retention of experienced, trained staff within the company. Additional benefits may include a shorter induction period, easier assessment of the candidate's potential and performance, and a greater flexibility as to the timing of actual job change. A possible disadvantage of making internal appointments is that the appointee may lack outside experience and fail to bring new ideas to the post.

The most common method of external advertising of technical vacancies is through the local, national or specialist trade press. It is important that the supervisor is involved in the design and placing of the advertisement, and they should be prepared to offer advice particularly where specialized posts are concerned.

● *Local Newspapers.* These offer a relatively cheap method of reaching a large number of local readers. There is usually considerable choice available covering the immediate locality, a wider region using a number of local papers in the same group, and daily, evening or weekly papers. The choice of day of the week in which the advertisement should appear is a matter of experience but many feel that if either a weekly or a daily paper is used, a Friday edition offers the best results. Local newspapers are most suitable for lower graded jobs and trainee posts where the salary may not be sufficient to warrant a potential employee moving from outside the area, or for those senior posts where there are a number of similar positions with other employers locally.

- *National Newspapers.* National newspaper advertising tends to be relatively expensive and as such is only suitable for senior posts. The choice of newspaper should be made with care after examining copies of each paper under consideration for at least a week. After this minimum research the selection of an appropriate paper should not be difficult.

- *Trade Newspapers and Specialized Magazines.* Covering as they do, a particular market, these can offer a very cost-effective means of reaching specialist and senior staff. A disadvantage of the laboratory magazine as against the better known scientific press is that the former are usually only published monthly whereas the latter appear weekly. However, the 'laboratory' papers probably receive wider circulation amongst technicians up to HNC level.

### The Advertisement

The advertisement needs to centre around three main factors, which are often summarized as I, We, You:

> I—is the identification of the organization
> We—the work of the organization
> You—details of the person required.

These three factors can be extended to nine items of information which are:

(i) Name of the organization, and any department concerned.
(ii) Job title and grade *e.g.* Scientific Officer, Laboratory Superintendent grade 7.
(iii) The main responsibilities and tasks involved *e.g.* supervision of seven technicians, provision of a technical service for research and teaching, design and construction of research apparatus.
(iv) Qualifications, work experience personal factors *e.g.* HNC, 10 years relevant laboratory experience, preferred age 30–35, full driving licence.
(v) Salary range, fringe and other benefits, *e.g.* company car, regular overtime.
(vi) Training and other opportunities, *e.g.* day release provided for further study.
(vii) Special concessions *etc.*, *e.g.* help with relocation expenses, discount on company products, share purchase scheme.
(viii) Procedure for applying *e.g.* application forms and further

particulars from; apply in writing enclosing a detailed curriculum vitae.

(ix) Company address together with the name and telephone number of the immediate supervisor who can supply further details and arrange informal visits.

It is useful to encourage potential applicants to seek further details of a job, such as a full job description, before they apply formally as this may discourage unsuitable applications and save the employer the time and effort of processing applications that may be withdrawn at a later date. An application form, if one is used by the organization, can be included with the further particulars.

### Application Forms

Not all organizations use formal application forms, preferring to rely on a letter of application and a curriculum vitae. The use of an application form does offer the advantage of ensuring that the information supplied by the candidate is adequate and provided in a format which makes comparison easier for those involved in the selection process. While the use of a standard form for all types and grades of staff may appear administratively convenient it is often not particularly useful to the candidate, as the type and extent of information required for a post as a senior scientist, bench technician, and porter may differ considerably.

Forms should be prepared for particular job groups, and as far as possible follow the same order as the Person Specification if one is used.

### SELECTION OF STAFF

A flow diagram showing stages in the selection process is shown in Figure 6.4.

### Preliminary Screening of Applications

Assuming an adequate number of applications are received, the first stage of the selection process should consist of preparing a 'short list' of about six candidates to be invited to interview. Some organizations prefer a longer short list of up to ten candidates, but this involves considerable time and expense at the interview stage.

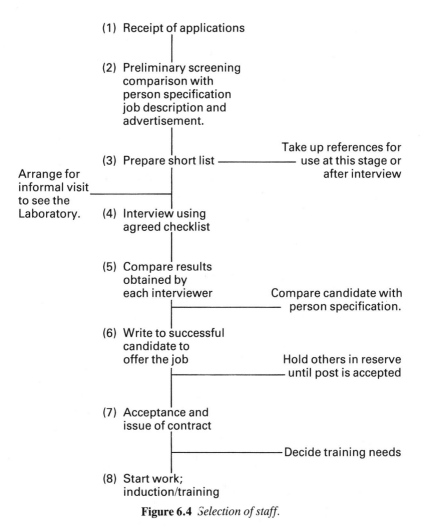

(1) Receipt of applications

(2) Preliminary screening
comparison with
person specification
job description and
advertisement.

(3) Prepare short list ——————— Take up references for
use at this stage or
after interview

Arrange for
informal visit
to see the
Laboratory.

(4) Interview using
agreed checklist

(5) Compare results
obtained by
each interviewer

Compare candidate with
person specification.

(6) Write to successful
candidate to
offer the job

Hold others in reserve
until post is accepted

(7) Acceptance and
issue of contract

Decide training needs

(8) Start work;
induction/training

**Figure 6.4** *Selection of staff.*

This preliminary screening, made on the basis of the Person Specification, may involve the technical supervisor; in some other organizations it will be done by the Personnel Department which will then forward the list to the section concerned. Obviously the former procedure is preferable as it enables the supervisor to get the 'feel' of the applications and avoids a good but unusual candidate being rejected because they do not fit the specification. This situation may arise when a candidate holds a qualification that, while being appropriate, is not on the approved list.

## References

Normally, references are requested in writing although some managers prefer verbal references in the belief that things may be said that would not be committed to print. Indeed it is not unusual for the applicant's present employer to be contacted on the 'old boy network' even if not given as a reference; a somewhat dubious practice. While references can be useful, they should be treated with caution, as employers wishing to retain a good employee may give a poor reference and conversely one wishing to get rid of a bad employee may give a glowing reference. References are best used as being supplementary to the main selection process and there are advantages in the interviewers not seeing them in advance of the interview.

## THE SELECTION INTERVIEW

The most frequently used method of selecting staff is that of the interview possibly supplemented by a medical examination. Additional tests are used by some employers and these will be discussed later.

### Preparation for the Interview

As the main objective of the employment interview is to exchange as much relevant information as possible with the candidates in order to assess their ability to perform the duties required, the interviewer should do everything possible to put the candidates at their ease and encourage them to talk.

The candidate will always be at a disadvantage during the interview as the interviewers are on their 'home ground', have the authority of their position and the trappings that go with it—secretary, desk *etc.* and, most important, have the job to offer. The interviewee is often a stranger, competing alone against others and will, if accepted, work under the interviewer. While the problem of disadvantage cannot be solved it can be reduced by giving consideration to the format of the interview and the physical surroundings.

### The Panel Interview

In the majority of public sector and related laboratories applicants for technical posts will be subjected to interview by a panel of between three and six people. This method of interviewing is traditional, allows the maximum number of people to be seen to be involved (often an important political point affecting status within the organization) and may reduce the effect of individual prejudice amongst those making the selection.

Amongst the disadvantages of the panel system is that it inhibits the candidate, increasing the feeling of disadvantage both in respect of being out-numbered during the interview, and the actual physical arrangement of the furniture in the room in which the candidate sits alone to face the united panel. The panel system is also very difficult to manage effectively without detailed briefing, particularly if certain members are present only because of their status rather than any direct involvement in the work of the post. The ineffectiveness of such a panel in providing a means of communication is increased by the fact that members of the panel may feel that they are competing against each other and are obliged to ask at least one question to justify their inclusion on the panel. Such an interview will provide little information for a systematic choice but the supervisor can ensure that his recommendation is made systematically by the use of one of the methods discussed later.

The panel system may actually reduce the influence of those who will be directly involved in the supervision or work of the candidate where the final selection is made on the basis of a vote by all the panel members or on the views of the most senior people present. Indeed, in some cases the immediate supervisor may be excluded from the formal interview panel, which would consist of a representative of the Personnel Section, a senior scientist or head of department, and the chief technician/scientific officer. In such situations the immediate supervisor may be invited to show the candidates round the laboratories and if this is the case, the opportunity should be taken to conduct informal interviews. If the supervisor is then asked for an opinion on the candidates this should be supported by the submission of a completed checklist or profile as a case made in writing is given greater credence than a mere verbal opinion.

**The One to One Interview**

The individual face-to-face interview reduces the feeling of disadvantage on the part of the candidate both in respect of equality in numbers and in the scope it offers for the use of less formality. This can be extended to the seating, possibly using an informal chair arrangement, or if an office must be used moving the chairs to the side of the desk rather than facing each other across it. McHenry ('ISTox Laboratory Management Symposium Proceedings 1979') suggests more detailed planning, for example ensuring freedom from interruptions such as telephone calls and placing a clock behind the interviewee so that the interviewer can keep a check on the time without the candidate being aware of, or distracted by, such monitoring. If the interviewer is adequately prepared this type of interview can result in much better communication between the two parties and consequently in a more informed decision.

The face-to-face system can be adapted to enable several members of staff to be involved by rotating the candidates to see several interviewers. This enables the people who would normally be involved in the interview panel to see each candidate while offering the advantages of a face-to-face interview. Used in conjunction with the methods discussed below it meets the needs for both maximum participation and a systematic choice.

## Conducting the Interview

The interview may be seen as an attempt to forecast the future behaviour and ability of a candidate based upon an assessment of the past and the characteristics that may be identified. The interviewer will want to gather information as to the precise details of the applicant's career, clarify points arising from the curriculum vitae or application form, discover the extent of their knowledge and interests in the subjects claimed on the form, and discover their likes, dislikes, and opinions.

The methods used to examine both candidates and application forms should be systematic rather than subjective. A number of procedures have been published to achieve these objectives. Professor Rodger of the National Institute of Industrial Psychology produced a seven point plan (Figure 6.5).

John Munro Fraser in 'Employment Interviewing' (1970), gave a five-fold framework using the following factors to separate people into five grades based upon the normal distribution of human differences; impact on others, qualifications and experience, innate abilities, motivation and emotional adjustment. A modification of this for use with technical staff is shown in Figure 6.6. This excludes both the lower levels of performance and those higher levels which are more appropriate to job groups other than those of technicians. Under the original framework proposed by Fraser the majority of technicians

| | | |
|---|---|---|
| (1) | Physical factors | physique, health, appearance, hearing, speech. |
| (2) | Attainments | educational and occupational. |
| (3) | General intelligence | both normally displayed and of which the candidate is capable. |
| (4) | Special aptitude | manual, mental, intellectual. |
| (5) | Interests | intellectual, practical, physical, artistic, social. |
| (6) | Disposition | acceptability to others, reliability and self-reliance. |
| (7) | Circumstances | domestic *etc.* |

**Figure 6.5** *The rodger seven factors.*

would have fallen into grade C (middle 40% of the population) and grade B (upper 20%).

The modified framework shown in Figure 6.6 is designed to cover the range of technical jobs and levels of attainment. As such it provides an aid to the preparation of person specifications and classifying applicants at interview.

Where a less complex system is preferred, or as a means of making notes during the interview for transfer to the framework at a later time, a simple checklist system can be most useful. Such a checklist is given in Figure 6.7. Yet another approach could involve the use of a scale of five or ten points for each quality, these can then be totalled and used to compare candidates.

## INTERVIEW STYLE AND APPROACH

There are four main stages to an employment interview.

### (1) Introduction

The first few minutes of the interview are all important in both setting the style of the interview and putting the candidate at ease.

One of two main approaches may be adopted:

● *Common Link or Shared Interest* is applicable to those occasions where the interviewer has been able to establish a shared interest or

| | A | B | C | D | E |
|---|---|---|---|---|---|
| **Impact on others:** | | | | | |
| (i) Appearance | Smartly and appropriately dressed | Well turned out | Undistinguished but tidy | Scruffy | Dirty, untidy |
| (ii) Voice | Pleasant to talk to, responsive | Good conversation and vocabulary | Limited conversation, good voice | Difficult accent | Difficult to understand, limited vocabulary |
| (iii) Manner | Attractive | Confident but not over-confident | Unremarkable | Lacking in confidence | Unpleasant |
| **Qualifications and Experience:** | | | | | |
| (i) General Education | University degree | 'A' Levels | Less than 4 'O' levels or CSE grade 1 in Science | CSE grades 2 or 3 | No formal education, or CSE at less than grade 3 |
| (ii) Vocational Education | FIST/FIMLS, MIBiol *etc.* | HNC/HND/CGLI FTC | National Certificate/ Diploma | CGLI craft cert. | No formal education |
| (iii) Work Experience | Managerial skills & related science | Supervisory posts responsible for day to day functions | Bench skills in appropriate subject | Limited experience relevant to post | No relevant work experience |
| **Innate abilities:** | Quick and active, able to interpret, develop | Capable of innovation and adaptation of ideas | Capable of a wide range of tasks | Capable of routine activities | Slow in understanding detailed tasks |
| **Motivation:** | High level of enthusiasm and self-realization | Hardworking, shows initiative | Adequately motivated for routine work | Lacks initiative, works well when busy | Poor, needs constant supervision |
| **Emotional Adjustment:** | Remains calm under stress, reliable and positive at all times | Accepted by group as leader, dependable | Capable of dealing with unforeseen and last minute problems | Fits into normal group in routine tasks | Tendency to be awkward or unhelpful |

**Figure 6.6** *A five-fold framework based on that of Fraser.*
(Modified after Betts to suit technician posts.)

| Name | Job Applied For: |
|---|---|
| **Age** | |

| Education *etc.* at School. | Further Education |
|---|---|
| CSE grade 2 or more | Mode of Study: |
| CSE grade 1 | National Cert. |
| O levels | HNC. |
| A levels | Post HNC. |
| Offices | Degrees *etc.* |
| Societies | Prizes |

| Activities: Professional | Social |
|---|---|
| Membership of Inst. | Services *e.g.* scouts *etc.* |
| Active in Inst/Assoc. | Clubs |

| Previous or other jobs | |
|---|---|
| P/T at school | Relevance of skills |
| Previous jobs | Responsibilities: |
| Reasons for leaving | for staff |
| | for money |

**Other Activities/hobbies**

| Physical Appearances *etc.* | |
|---|---|
| Dress/turn out | Confidence +/− |
| Voice | Disabilities |

**Preference for practical or academic work**
**Prepared to change area/move**
**Ability of candidate to fit in well with existing staff**
**Amount of training required to perform the job**

**Figure 6.7** *Interview checklist.*

other link with the candidate on the basis of the application form or curriculum vitae. This may be related to a geographical area, school, previous employer or a mutual interest in a sport or hobby. A brief discussion on this link may help the candidates to relax and establish a rapport.

● *The Off Balance approach* requires considerable expertise and confidence on the part of the interviewer if it is to be used successfully.

It consists of the interviewer welcoming the candidate and then discussing something that has no obvious connection with the interview. This approach is only applicable to face-to-face interviews while the common link can be used by the chairman of a panel.

## (2) Questions

During the interview the interviewer should lead the candidate to talk about the points the interviewer wishes to establish, based upon the checklist. Where necessary the candidate should be guided back onto the required subjects but without interrupting the flow of conversation. The interviewer should avoid the use of questions that can be answered with a monosyllable. The use of open questions should help to ensure that the candidate does more talking than the interviewer. Indirect questions may be used to discover the candidate's views and opinions rather than the factual reply that might arise from direct questions. For instance if asked 'Did you enjoy working for your present employer?' the reply will almost certainly be 'Yes, but . . .', while in response to the question 'Why are you leaving?', the standard answer is 'To further my career prospects'. If then asked 'What is your opinion of the organizational structure and manpower planning policy of your present employer?' the candidate is more likely to give an answer that informs the interviewer of some of his/her attitudes and allows scope for follow up questions.

## (3) Informing

The third stage of an interview consists of the interviewer providing the candidate with more information relating to the terms and conditions, job content *etc.* than has been possible before the interview. It is also useful to explain the organization's manpower policy as it will affect the candidate, promotion prospects, and organizational structure. The candidate should be encouraged to ask questions on these topics, with the interviewer briefly summarizing the main points if there seems any doubt or confusion. Agreement should be reached on a starting date should the candidate be offered the post.

## (4) Concluding the Interview

The candidate should be thanked for attending and be given details of how to claim a refund of interview expenses. A date should be given as to when the candidate may expect a reply to the application. Details of the acceptance procedure should also be outlined.

## Assessing the Candidate

Each interviewer should assess the information obtained from the interviews as an individual, comparing the impressions obtained with the person specification. The candidates should be arranged in order of preference prior to discussion with the other interviewers.

If there are differences between the first choice of the different interviewers, the interview checklists and the relative importance of differences between the interviewers' findings on specific items should be used as the basis of the final selection. In cases where the interviewers have differing results, the tendency to make the final choice on the basis of the overall impression created by the candidate runs counter to the attempt to introduce a more systematic approach to the interviewing process and should be avoided.

The offer of appointment should be made to the successful candidate as soon as possible after the interview. The letter should contain, in addition to the time and date upon which the candidate should start, the name of the person to whom the new employee should report.

The letter should contain a request that the candidate indicates acceptance of the offer immediately. As soon as this acceptance has been received the other applicants should be notified that they have been unsuccessful.

The letter of appointment may be worded in such a way as to provide a Statement of Employment under the Employment Protection (Consolidation) Act 1978. Alternatively a statement of employment may be included with the letter to avoid the need to issue one later.

If none of the candidates prove acceptable, letters of rejection can be sent to each of them immediately.

## Staff Selection Tests

An increasing number of employers are using intelligence, aptitude or personality tests as part of the selection process. However, the design, use, and interpretation of these tests is a matter for experts and as such they are not appropriate for use by the supervisor. One type of test has been used by line supervisors for many years, although not necessarily in the technical sphere. This is the Work Test, where a candidate is given a test to check the standard of a claimed skill. Such tests are routinely given to typists to check speed and accuracy and there is no reason why specific bench skills relevant to the post could not be assessed during the selection process. This might be particularly useful where a candidate is applying for a post from full-time education *i.e.* with a degree, HND or a National Certificate based on full-time rather than part-time attendance.

# CHAPTER 7

# Salaries and Gradings

## SALARIES

The majority of technicians appointed by large organizations will be paid a monthly salary based upon salary scales negotiated nationally between the employer, or a group of employers, and the relevant trades unions.

There are a number of other payment systems that may be encountered by the technical supervisor.

### Hourly or Weekly Rate

This provides a system that is simple to operate but offers no financial incentive for the employee to be more productive as payment is based on attendance not output. The wage for the basic week is agreed and divided by the number of hours worked to give the hourly rate.

Attendance in excess of the normal working day is paid at enhanced (overtime) rates either in money or time off in lieu.

### Piece Work Systems

These are normally only applicable to manual workers and are based on payment for each piece of work produced. The payment is calculated by timing the unit of work at a reasonable speed. The system provides a financial incentive linked to output but may require detailed quality control systems. Careful monitoring of safety precautions is also necessary as there may be a temptation to 'cut corners' in order to increase payment.

### Measured Day Work

This is a system of payment by results which may be applicable to technical staff in production industries. Under such systems a bonus based on performance may be paid in addition to the normal wage. In the simpler schemes the worker is given a fixed rate of pay for reaching an agreed output. This latter system although easy to administer offers no incentive for the worker to increase output above the agreed level.

## Bonus Schemes

Under group bonus schemes the group rather than the individual is paid for increased output. It is argued that such schemes benefit and encourage the group to work as a team. Much of the success of these systems depends on the size and compatibility of the group. Supervisors responsible for staff working group bonus schemes often share in the bonus in what are known as Indirect Labour Bonuses.

## Profit Sharing Schemes

Such schemes are considered to encourage additional effort on the part of staff and link an additional payment, over and above the normal salary, to the performance of the company. Unlike bonus schemes the additional payments are normally paid annually. Share purchase schemes, or the issue of free shares, are a variation on this theme.

## Fixed Annual Salaries

A number of technicians within industry will be paid a fixed annual salary. This salary may be open to annual review or only be reviewed on the basis of an application by the job holder. Such reviews normally consist of two elements, although they may not be identified as such, a cost of living award and a merit award based on the individual's performance.

A variation on this system is for the employer to have salary scales for each grade but with technicians being held at a fixed point within the scale. When staff are paid on fixed points the employer often regards the actual salary as confidential, not favouring comparisons between individuals.

## Incremental Salary Scales

This system is used for technical staff throughout the public sector and in many large private companies. Jobs are graded with each grade having a series of incremental points. The grade in which an individual is placed will depend on job content and/or qualification but the position in the grade will depend on length of service.

There may be considerable overlap between grades which will necessitate a system to provide an appropriate salary increase upon promotion to another grade. The disadvantage of this system is that it rewards length of service not performance. The use of additional increments within, or above, the grade as merit awards for particularly good performance may counter this. In addition to each member of staff receiving an annual increment until the top of the scale is reached, a cost of living award normally results in revision of the scales annually. Where short incremental scales are used resulting in a large proportion of staff being on the maximum salary there will be pressures for larger cost of living increments, promotion with the danger of 'grade drift', or loss of staff.

**Overtime Payment**

In addition to the hourly rate system of payment a number of the other methods discussed allow for the payment of overtime by calculating the annual salary at an hourly rate for this specific purpose.

Where the organization has a policy of payment for overtime, rather than time in lieu, the supervisor will have a duty to ensure that such payments remain within the amount allowed in the budget. If necessary the approval of the head of department or finance section should be obtained before a regular commitment to overtime is allowed.

Where additional hours worked are awarded by time off in lieu the responsibility for authorizing the work should rest solely with the supervisor who should ensure that such time off does not reduce the effectiveness of the unit. Approval of holidays and other time off should also rest with the supervisor as the person responsible for the group.

## JOB EVALUATION AND GRADING

Many scientific and research laboratories operate job evaluated grading systems for their technical staff and it is essential for the supervisor to understand the method of evaluation both as a line manager controlling staff and as an individual offering support and advice to subordinates seeking advancement.

The importance of recognizing and rewarding staff for their contribution is discussed elsewhere in this book. Job evaluation is regarded as a means by which this recognition may be achieved in terms of money and status. These two items are viewed by employees in two ways. At the *general level* they seek a salary which will support the living standards they consider appropriate to their level in life. In addition they have a desire to ensure that their income and status is fair with reference to other employees. It is this *relative position* of salaries and status that may result in dissatisfaction amongst staff affecting motivation and performance.

**Job Evaluation**

Job evaluation was introduced to solve the problem of pay inequalities by determining the relative worth of jobs. It is based upon the assumption that the most important factor in determining salary is job content rather than factors such as personal abilities, excellence of performance, length of service or indeed qualifications. In practice, allowance may be made for these items either to gain entry to a grade or to receive an additional increment within it.

The purpose of job evaluation is to order jobs through logical and systematic comparisons of job content. Unfortunately, in fact it is often job descriptions that are compared (see Chapter 6) and in practice these may differ from the actual job performed.

Job evaluation has a long history with one of the first attempts being introduced by the US Civil Service Commission in 1909. In 1924 Merill R. Lott produced a method based on fifteen factors designed to measure skill, effort, responsibility, working conditions and other items now considered to be extraneous such as the size of the organization.

The use of the technique has grown over the years until, by 1969 (when technicians in universities were first evaluated by this method), approximately a quarter of all employees in Britain had pay levels assessed by this method.

There are two main systems of job evaluation.

● *Non-analytical*. In the non-analytical methods of job evaluation, appropriate for use only when jobs have something in common, jobs as a whole are compared. The method fails to quantify differentials between jobs.

● *Analytical*. This is a more complex method which defines and identifies the common components (factors) in jobs, *e.g.* knowledge, responsibility, complexity, accuracy, hazards *etc*. This allows comparisons to be made between jobs which superficially have little resemblance.

Within the technical sphere one of the largest job evaluation exercises was undertaken by the Manpower Productivity Service (MPS) in fourteen universities, covering technical staff in laboratories, electronic and electrical workshops, mechanical engineering, animal houses, photographic and specialist sections. The grading structure resulting from the survey by MPS of 2952 jobs was adopted by those universities party to the agreement and used with some modifications by local authorities. As this structure was so widely adopted it is used here to illustrate the technique of evaluating and grading technical jobs.

### The Factor Plan

A nine factor plan (Figure 7.1) was used to discriminate between the jobs undertaken by the technical staff. The use of so many factors is unusual as normally only three or four factors are used, so as to avoid one particular facet of a job being counted under different factors *e.g.* under complexity and accuracy.

| Factor | Number of levels | Points per level | Maximum Weighting |
|---|---|---|---|
| Education | 4 | 2 | 8 |
| Experience | 5 | 4 | 20 |
| Supervision received | 4 | 4 | 16 |
| Supervision given | 6 | 3 | 18 |
| Relationships | 4 | 4 | 16 |
| Job complexity | 4 | 3 | 12 |
| Accuracy | 3 | 2 | 6 |
| Mental strain | 3 | 1 | 3 |
| Hazards | 3 | 1 | 3 |
|  |  |  | Total 102 |

**Figure 7.1** *University technical staff job evaluation factor plan; showing the relationship between the various factors.*

The MPS survey indicated that there was a wide range of salary levels attached to jobs with similar factor plan scores, with clusters at the entry and exit points of the Technician and Senior Technician scales (Figure 7.2).

As a result of the job evaluation exercise, a new grading structure was introduced using twelve grades (subsequently reduced to ten by combining those at the lower levels (Figure 7.3) ).

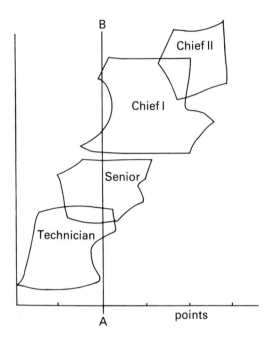

**Figure 7.2** *Results of analysis of university technician jobs prior to restructuring, illustrating overlap between grades with jobs at the line A–B being performed by staff paid as technicians, senior technicians and chief technicians.*

### Implementation of New Grading Structure—'Benchmarks'

Implementation of a new grading structure obviously produces a number of difficulties particularly when generalized benchmark job descriptions are used to slot people into an appropriate grade rather than preparing detailed job descriptions and factor plans for everybody. The supervisor and the individual should normally be involved at this stage to ensure accuracy and to avoid subsequent appeal against the grading.

### Appeals Machinery

All regrading systems should contain provision for the job holder to agree to the job description as a true record and to appeal against any grade awarded. The appeals machinery may use the grievance pro-

| Grade | Points | Jobs |
|---|---|---|
| 1A | Under 30 | Stores, laboratory, workshop and animal house |
| 1B* | | assistants |
| 2A* | 31–35 | Animal house, stores and reprographic technicians. |
| 2B* | 36–40 | Teaching laboratory, photographic technicians. |
| 3* | 41–46 | Research laboratory, specialist services, mechanical engineering and electronics technicians. |
| 4 | 47–52 | Animal house, stores, teaching laboratory and photographic (senior) technicians. |
| 5 | 53–59 | Research laboratory, specialist services, mechanical engineering and electronics (senior) technicians. |
| 6 | 60–69 | Supervisors of laboratories and sections of between 3–10 technicians. |
| 7 | 70–79 | Supervisor responsible to Head of Department in small department of up to 12 technicians or responsible to grade 8 in large department. |
| 8 | over 80 | Supervisors responsible for a complete technical service covering teaching, research, workshops and supporting services. Between 13–30 technical staff. |

* Subsequently combined to make new grade 2 (1B + 2A) and 3 (2B + 3).

**Figure 7.3** *University technical staff grading structure after job evaluation. It will be seen that at the non-supervisory level job progression was normally through alternative grades (i.e. odd or even numbers) avoiding overlap and resulting in an appreciable salary increase upon promotion.*

cedure or be a separate panel, usually consisting of representatives of both management and unions.

Job evaluation is not the answer to every problem relating to salary and status nor does it establish whether the jobs examined are constructed in the most satisfactory way. In using any grading system based on job content provision should be made for recognizing the career requirements, as well as educational and technical efforts of individuals.

Grading systems, even job evaluated systems, should not be considered as sacrosanct. Technology changes, salaries are eroded by inflation, and grade drift will occur as individuals and managers learn how to use the system to satisfy their own vested interests. After a structure has been in use for perhaps ten years the supervisor will be faced with the problem of whether to keep rigidly to the grades as defined or to seek maximum rewards for subordinate staff even if this risks undermining the published structure. Obviously if grade drift takes place in other sections, while the supervisor of one laboratory adheres to the grade definition, this will adversely affect the motivation and turn-over of staff in that particular group.

The problem of grade drift should not be under-emphasised. As the drift is always upwards, the result can be that of all 'chiefs and no indians'. This will either mean that no-one wants to do the basic work as it is not their job, or that staff continue to do the same manual tasks but do not show them on their job description.

In some organizations a more rigid organizational structure is applied which prevents grade drift and ensures that the 'establishment' is maintained at a level appropriate to the work and objectives of the organization.

Within the Scientific Civil Service, staff who consider themselves under-rated may apply for promotion. If, following investigation and interview the application is approved, the individual is offered a post at a suitable grade elsewhere within the organization and their original post is filled by someone else.

# Induction and Monitoring of Staff

The induction of new staff is often considered to be part of the training process and this tends to lead to a degree of confusion about the function of both. Training may be defined as the acquisition of skills, technical, managerial, or social, while induction introduces new employees to their jobs, the organization and colleagues. All new staff need induction, not all need skills training.

Induction should be designed to assist the new member of staff to settle into the job as quickly as possible with the minimum of stress. It should also help in the development of a positive attitude towards the organization, acting as the second opportunity to motivate new staff (the selection interview having provided the first).

(1) Introduction to the organization
(2) Products and services
(3) Conditions of service/legal aspects
(4) Rules and disciplinary procedures
(5) Trades unions
(6) Organizational structure of section
(7) Health and safety
(8) Social facilities
(9) Tour of site

**Figure 8.1** *The induction process.*

Figure 8.1 shows nine factors of an induction programme. It is not necessary or desirable to cover all of the items on the first day or even within the first week of employment. In addition, certain aspects are applicable not only when an employee joins the company but whenever a member of staff changes section, shift or department. The structure of the induction programme will vary with the size of the establishment and the frequency with which new staff are employed. In a small unit the process may be very informal on a one to one basis; in a larger organization it may be possible for at least part of the process to be organized for groups of ten or twelve new employees.

The induction process is, where a conventional management structure exists, part of the personnel function with the supervisor providing a complementary role in respect of the section or unit. However, despite workers such as Gormersall and Myers in 'Breakthrough in On-The-Job-Training' (*Harvard Business Review* 1966 **44** No. 4.) reporting that an effective orientation programme is very cost effective, in many laboratories there may not be such a system. In such instances the personnel officer will cover the legal aspects and provide details of terms and conditions of employment but no provision will be made for the other factors. Under such circumstances the technical supervisor should assume responsibility for devising and operating the induction of staff. In such cases those factors relating to the overall organization (primary induction) and those to the section and the particular job (secondary induction) may be combined.

## INDUCTION PROCEDURES

### Primary Induction—Personnel Section

*Introduction to the Organization.* The welcome to the employee and a brief introduction to the company, its aims and objectives followed by the provision of brief details of the overall organizational structure, types of staff employed and the relationship of new employee's unit or section within the organization.

*The Products and Services of the Company.* This section is relatively straight-forward where the company is involved in the manufacture or testing of a product, or in educating students. Where the organization is concerned with academic research it may be slightly more difficult to be specific but it is important that sufficient information is given to impart a degree of enthusiasm.

*Conditions of Service.* The employee must be issued with a Statement of Employment within thirteen weeks of commencing work giving basic details of employment. This will be considered in greater detail later under Employment Law. During the induction, the hours of work, holidays, and pay should be explained. This section should include details of salary scales if they are used and the procedures for reviewing salaries. If individual salaries are regarded as confidential this should be made clear to the employee.

Details of the pension scheme should be provided both in respect of payments and benefits.

*Rules and Disciplinary Procedures.* It is important that staff are made aware of company rules before they infringe them! During the induction process such rules should be explained as should the disciplinary and grievance procedures.

*Trades Unions.* The company policy towards trades unions and staff associations should be explained as appropriate. If a union is recognized for bargaining purposes this should be indicated. Union involvement in the organization, disciplinary, grading and safety machinery should be explained.

## Secondary Induction and Supervision

*Organizational Structure and the Work of the Section.* This is best undertaken by the section or line supervisor. It should include an explanation of the function of the section and of the staff structure. The role and duties of the new employee should be explained with reference to the job description and the work of the other members of the team. The newcomer should then be introduced to the other members of the section and to 'a sponsor'.

The sponsor system consists of placing the new employee in the care of an experienced technician who will be responsible for helping in familiarization with the job, for assisting the new member to fit into the section, and for taking them to coffee, lunch *etc.* It is probably best if the sponsor is made responsible for the issue of protective clothing and explaining safety procedures. This will help to ensure that safety is considered as an integral part of every technique rather than a theoretical topic.

*Health and Safety.* A preliminary introduction to health and safety may be given by the personnel department but it would seem to be given more relevance if it is dealt with at the secondary level.

The supervisor should explain the organizational aspects of the company's safety policy (briefly at this stage); further details of safety committees *etc.* may be given at a later date. During induction emphasis should be on ensuring that the new employee understands sufficient practical details to work safely, together with alarm and escape procedures. Copies of safety rules, Codes of Practice, *etc.*, should be issued at this stage. In certain laboratories it may be necessary for staff to receive a degree of medical monitoring *e.g.* X-ray or immunization. The requirement may be greater in other laboratories when staff will be registered as Radiation or Biological workers and subjected to more detailed screening. Such registrations are best made early in the induction process.

*Social Facilities.* Provision of details of any Social Club, sports and refreshment facilities.

*Tour of Department and Site.* This should be in two parts. The first part, a tour of the building in which the new employee will work, should include the situation of eating facilities, toilets and safety items *e.g.* fire extinguishers, escape routes, assembly points, first aid boxes, emergency showers, alarm points.

The second stage of the tour, which may be at a later date, should consist of a walk round the whole site if it is not too large. This should include, in addition to other laboratories, central stores, offices, the social and recreational facilities, and alternative car parks *etc.*

*Review of Progress.* The supervisor should, as part of the induction process, meet the new employee at the end of the first day to discuss progress and impressions.

At the end of the first week the supervisor should check that the new employee has assimilated the main points of the induction process and discuss his progress with the sponsor. The discussion with the new-comer should provide feedback on progress and any areas where performance is less than was hoped for at this stage. This may indicate a training need and part of the monitoring process should provide an opportunity to decide on necessary training. Most employees will be appointed for a probationary period, perhaps three months, and the supervisor should monitor progress throughout this period. It is important that the newcomer is kept informed as to whether progress is satisfactory. Often only poor performance is discussed with the employee and lack of feedback may make the satisfactory employee anxious. At the end of the probationary period the supervisor should ensure that the new person has been accepted as a permanent em-ployee and a member of the team.

## MONITORING OF PERFORMANCE

The monitoring of staff, and resultant feedback of results, is recog-nized as an essential part of any formal training programme and to a lesser extent as being applicable to new staff. In many laboratories these are the only circumstances in which appraisals, involving the member of staff, are routinely used. Counselling interviews during informal and formal warnings are incorporated in most disciplinary systems but these procedures are only invoked after a problem has been identified, and as such do not form part of the monitoring of performance, and are essentially of a negative nature. The regular use of structured appraisal systems offer a positive approach and may help to avoid the need for such actions. Appraisals should improve communication within the laboratory and by demonstrating that management has an interest in staff, contribute to the maintenance of motivation (Figure 8.2).

Regular appraisal offers the supervisor the means of identifying:
  (1) Present and future training needs.
  (2) Areas where performance could be improved both in respect to techniques and staff.
  (3) Members of staff working at less than full potential suitable for promotion or job enlargement.
  (4) Hazards and difficulties in the work before they become major problems.

If handled sympathetically by the supervisor the system offers subordinate staff:
  (1) Recognition for the efforts and achievements, even if this cannot be in terms of financial reward.
  (2) Information on career and promotion prospects. (It is important that staff are shown that management are concerned for the future of their staff.
  (3) Information on the future of the group, plans, prospects and reasons for developments.

**Figure 8.2** *The benefits of the appraisal system for supervisor and subordinate.*

In those organizations where appraisal systems are not used, there is no reason why the supervisor could not introduce them within his group on an informal basis. Such appraisals would need to be introduced with care so as not to raise objections from subordinate staff. This may be done by placing emphasis on the positive aspects of the individual's performance, and by using the meeting to discuss future career development. Such an innovation could be used to motivate subordinate staff and increase the influence of the supervisor if the results were submitted to management to coincide with an annual review of grades of salaries.

There are four main systems of staff appraisal used by industry of which only one, the appraisal interview, is appropriate for instigation by the supervisor. The use of performance ranking, rating systems and essay reports, which are more formal in nature, would need the support and participation of management.

## THE APPRAISAL INTERVIEW

The actual format of the interview will vary with the system of appraisal used.

### Supervisor Initiated Schemes

Where the monitoring system has been introduced by the supervisor and the interview consists of a discussion between the supervisor and a subordinate, the style of the interview will be less formal than in the systems discussed below where the interviewer is a member of management. For the supervisor/subordinate interview to be constructive the supervisor will need to prepare an outline for discussion in advance and ensure that the initiative is not lost during the discussion.

The employee's job description can form the basis of the initial part of the interview. This will provide a structure for discussing the progress of the subordinate and for detecting any changes that might have occurred in either the nature or ratio of the tasks performed. If the employee is attending a course of study the college report can also provide a useful basis for discussion. In both cases the supervisor should encourage the interviewee to do most of the talking, to discuss any weaknesses, and to identify areas of concern. Self-criticism is more acceptable than criticism from others and should lead to suggestions from the subordinate as to means of correcting faults. The interviewer can at this stage introduce any criticisms that he has of the employee's performance, preferably in agreeing with the self-criticism and making constructive suggestions as to possible improvements.

Before concluding the interview, the future of the subordinate's job and career developments should be discussed. This might entail attendance at college courses, further training or, if future advancement is limited, an honest discussion on the merits of changing jobs, either within or outside the organization.

At the end of the interview, the supervisor should sum up the discussion and ensure that there is a clear understanding of any decisions reached.

After the interview, the supervisor should record the points agreed in the summary and a report can then be submitted to management based on all of the staff summaries.

## Management Initiated Appraisal Interview

Where management has introduced a system of appraisal interviews they are frequently conducted by someone other than the employee's immediate supervisor and are often used in conjunction with a report prepared by the supervisor.

In those organizations such as the Civil Service where a 'grandfather' system of appraisal interviews operates, the employee's immediate supervisor completes an appraisal report form on the individual while the interview is conducted by the supervisor's own immediate supervisor. Such a system is said to improve communication within the section, offer support to the inexperienced supervisor, and provide a means of monitoring the attitudes of supervisors towards their staff.

In other organizations a member of the personnel section may conduct the interview but as they have less technical knowledge than the line supervisors the emphasis of the interview tends to be one way.

There are a number of variations on this type of appraisal system. In some cases the supervisor's report is regarded as confidential and it is not seen by the subordinate, in others it is seen and initialled (but not changed) by the subordinate. In a third system, differences of view are noted, and in a fourth the report must be jointly agreed.

## Ranking Systems of Appraisal

In general, ranking systems are more appropriate for use on the factory floor than in laboratories as, in order to be operated effectively, they require a large number of staff. The system is also very subjective and, as the ranking does not have to be supported by documentary evidence, is difficult to evaluate.

The simplest ranking system consists of listing all staff in order of performance or merit. More complicated systems entail placing staff in grades ranging from exceptional to poor.

## Rating Systems

Rating systems are designed to indicate the employee's strengths and weaknesses and are a considerable improvement on ranking procedures. An example of a rating system appropriate for use with technicians is given in Figure 8.3. There are a number of rating scales ranging from 'good', 'satisfactory', 'poor', to six point scales which

Name:                          Job Title:                  Grade:

Qualification:                 Age:                        Time in grade:

Reviewed by:

|  | Satisfactory | | | Unsatisfactory | |
|  | v. good | good | average | poor | v. poor |
|---|---|---|---|---|---|
| Knowledge of job/expertise | | | | | |
| Quantity of work | | | | | |
| Accuracy of work | | | | | |
| Adaptability | | | | | |
| Dependability | | | | | |
| Attitude to safety | | | | | |
| Use of initiative | | | | | |
| Performance on educational training courses | | | | | |
| Attitude on educational training courses | | | | | |
| Motivation | | | | | |
| Communicative ability | | | | | |
| Ability to cooperate with others | | | | | |
| Relationships with: subordinates peers supervisors scientific/academic staff | | | | | |
| Organizing ability | | | | | |
| Willingness to take responsibility | | | | | |
| Leadership skills | | | | | |
| Suitability for promotion | | | | | |
| Action to be taken | | | | | |

**Figure 8.3** *Appraisal rating form for technical staff allowing for three satisfactory, and two unsatisfactory levels of performance.*

include variations allowing for length of time in the job to be recorded. For the collection of such detailed information to be of value the organization must be prepared to act on it in terms of promotion, job enlargement and training. In practice, few laboratories have the flexibility or the expertise to do so.

As a result of the appraisal and monitoring of staff the supervisor should be aware of those members of staff deserving promotion or additional salary increases. If each supervisor has authority to make such recommendations, it enhances their position, and therefore the line structure of the organization.

**Salary Increases**

In organizations where staff are paid on annual incremental scales provision often exists for the award of an extra increment within, or on top of the scale, based on good performance. This 'merit' increase may be linked with the formal appraisal system or based on the supervisor's recommendation to the Head of Department.

Where staff are paid on personally negotiated salaries or on fixed points within a scale, increases may be made on, or supported by, the supervisor's recommendation.

**PROMOTION**

**Up-grading**

In organizations operating job evaluated grading systems there may be a provision for staff to apply for up-grading to another grade on the basis of changes in job content. There are two aspects of such procedures which affect the supervisor.

The first is the supervisor's responsibility to control and plan (see Chapters 1 and 2). The supervisor has a duty to management to see that changes do not occur within his section that are not part of the manpower or financial plan. Changes in structure resulting in employee initiated up-gradings may be at variance with both of these, may reduce the ability of the section to finance those promotions that it wishes to put into effect, and undermine the position of the supervisor, who is in practice by-passed by the individual. Within the technical structure of many organizations there is an additional complication in that specific qualifications are required for each grade. Where up-grading is sought solely on the grounds of change in job content this may undermine the 'qualified service' aspect of the structure. On the other hand it would be blatantly unfair if an unqualified member of staff was seen to be carrying out the duties of a certain grade but not being paid the 'rate for the job'. All changes in job-content should be under the control of the supervisor who should, if necessary, have the proposed duties added to the existing job description and the post examined by the grading panel to assess the effect on the grading before allocating the duties. In this way changes may be incorporated into the organizational plan and perhaps more importantly, the supervisor's role is enhanced.

## Conventional Promotion

Promotion, as distinct from up-grading, entails an individual changing jobs within the organization with an increase in duties and rewards.

The appraisal system should enable the supervisor and his manager to identify staff suitable for promotion. If a suitable post is available within the department the situation is relatively simple. The individual may either be offered the post or if necessary invited to apply when it is advertised. If there is no suitable position within the section the central Personnel Office may be able to arrange an inter-departmental promotion. Inter-departmental transfers for promotion purposes should be encouraged by supervisors and not regarded as a method of poaching staff. Such systems avoid the organization losing expensively trained people. Planned, or at least supported, transfers enhance the career prospects for staff thus improving morale.

In encouraging the career development and promotion of deserving staff, the supervisor will be literally left with those staff who have not been promoted, either because they have risen to their maximum level of achievement or have not yet developed sufficiently in technical or social skills. If these staff accept the situation there will be no problem, but if they do not, the supervisor will be required to use the appropriate counselling and/or training skills to maintain motivation.

## Transfers

There will be a number of cases where the appraisal system indicates that a member of staff is not performing satisfactorily. In such instances there will be a need for counselling and retraining before consideration is given to the possibility of reducing the job content or transferring the individual to another post. This is particularly the case where the poor performance is a result of ill health or personal problems rather than of bad conduct.

If the problem is of a short term nature and a change in job content could be achieved without affecting the individual's grading the supervisor may feel justified in taking action informally. In all other cases the Personnel Section should be notified and the organization's procedures for dealing with disciplinary matters should be followed to the letter, so as to avoid procedural difficulties at a later stage.

# Training

## INTRODUCTION

The organization and planning of staff training and development is normally considered to be part of the Personnel Function of the organization, with the line supervisor being involved at the sharp end by providing on-the-job skills and experience. Many existing books on Management and Supervision seem to assume the presence of efficient training sections within the personnel department and the employment of training specialists who can bring their particular skills and expertise to bear on training problems. Donnithorne in 'Essentials of Supervisory Management' (Ed. C. H. John, Cassell Ltd. 1983) writes that 'the responsibility for training or education remains with line management, but as with recruitment, Personnel are able to deal with the control and co-ordination of training needs'.

In our experience most laboratories do not have the benefit of such services. Indeed the 'Istox Laboratory Management Survey' (Institute of Science Technology, London, 1978) indicated that the Personnel Section was only involved in the organization of training (as distinct from educational courses) in 10–14% of laboratories, in contrast to 36% of laboratories where a technician was designated Training Officer, 29% where the technician prepared the training programme and 34% in which technical staff selected the actual procedures taught. The Personnel Section provided the actual training of technical staff in only 4% of laboratories.

The importance of the role of the technical supervisor in technician training often appears to be overlooked, possibly due to the emphasis placed on attendance at college courses to obtain formal qualifications and the restriction of 'formal' training to the 'trainee' or 'junior' grades. Both of these aspects give rise to considerable concern and one of the professional bodies for technicians, the Institute of Science Technology, has attempted to improve the situation by the introduction of an In-Service Training Scheme for junior staff, and the Business and Technician Education Council's continuing education programme for Laboratory Supervisors leading to Fellowship of the Institute. Training must not be seen as a discrete or separate area applicable only to certain grades of staff *e.g.* trainees, but an important and continuous process to include new staff, up-dating in new techniques for existing staff, improving performance, preparation for promotion, meeting the challenge of organizational or political changes, or to prepare for retirement.

The topic of training needs will be discussed later but first the differences between education and training and the role of the colleges in the development of technical staff will be considered.

The purposes of training most frequently given in books dealing with supervision and management relate to improving productivity, quality, and safety. The reduction in staff turnover resulting from effective training is also mentioned but on the whole it is clear, that while training is given to people, the object of the exercise relates to the product. This is often overlooked in the non-industrial sphere where the lack of organizational objectives is never clearer than in the attitude to training.

In part, the historical background may also be responsible for the present situation. The Industrial Training Act of 1964, although far from perfect, did result in the acceptance of the need for formal training within the industrial sector, and as a result, training came to be seen as part of the overall manpower plan of the organization.

The educational sector not only lacked the financial discipline of the private sector, it was excluded from the Act and was able to continue with the traditional approach to training. This resulted in the recruitment of staff against specific posts, the use of informal *i.e.* not organized, training and the absence of any national standards. Within the technical sphere attendance at a college day release course to obtain a professional or technical qualification was frequently considered to be all the training that was required to equip young people for their future careers. Such training as was made available was often a result of the actions of an enthusiastic supervisory technician with a sense of duty towards new recruits to their profession.

Training for promotion and the training of supervisors tended to be neglected or ignored completely, which naturally influenced the styles and attitude of staff in supervisory posts. This was particularly ironic as colleges, polytechnics and universities frequently offered off-the-job training for the employees of industrial organizations and were active in producing text-books on how to train others while neglecting their own staff. A useful approach is suggested by Boydell in 'A Guide to the Job Analysis' (BACIE, 1973) who defines systematic training as 'planning to give people the chance to learn to achieve the results that the job demands.'

The distinction between education and training is of some importance for, although closely related, they do require a difference of approach. For the purpose of this chapter the following definitions will be adopted:

• *Education.* Studies to obtain a qualification or knowledge of mainly theoretical nature, even though the topic under study may be of a technical nature, or where the subject is wider than that immediately required by the employing organization.

• *Training.* An approach concerned with the acquisition of practical skills but requiring sufficient theory to be imparted to enable the technique to be understood. In the context of training, practical skills are those acts which require practice before they can be performed adequately. The trainee can be told how to do these tasks at college but will require 'hands-on' experience before they can be performed to the standard required. Training should be provided so as to enable the employee to meet the aims and objectives of the employer. It is an activity which is the direct responsibility of the employer, although it may be contracted to a third party.

These definitions avoid attempting to classify education and training by the situation where they occur, *i.e.* college and work respectively, as the two may be interchangeable. They also make it clear that the two are complementary and as such should be co-ordinated. This in itself will require a considerable change in the attitudes currently held by colleges 'educating' and organizations 'employing' science technicians.

Before considering what should be the basis of all training, the analytical approach, and the role of the supervisor towards training, present systems of training and recent developments in this area will be looked at in more detail.

## TRAINING OF SCHOOL LEAVERS

Traditionally, formal training within laboratories has been confined to school leavers entering the technical service with GCE 'O' or 'A' levels at Trainee, Junior or equivalent grades.

Other new recruits whether young graduates straight from university with little or no bench experience, or technicians recruited from other employers unfamiliar with the procedures used within the new organization, are assumed to be able to acquire the necessary technical skills without resorting to an organized programme of training. Educational and Research Laboratories did not come under the aegis of the statutory Industrial Training Boards although some employers used training programmes based upon those of the Chemical and Allied Products ITB for laboratory staff, and some workshop technicians followed the recommendations of the Engineering Board. Apart from this a systematic approach was rarely adopted.

Most young technicians entering technical employment were recruited against specific posts and received on-the-job training to enable them to fulfill the requirements of that post as envisaged by their immediate supervisor. The training that was provided was usually unstructured and followed the procedure known as 'sitting next to Nellie'. Nellie being the senior technician or scientist, neither of whom will have received training in supervisory skills. The actual training, even when it met the requirements of the specific job, often failed to meet the trainees' career needs or provide the necessary bench skills to support the subjects being studied as part of the day-release programme at college. This latter factor is particularly important as students may study units as part of their BTec programme which would not normally be covered during job specific training at work *e.g.* Physical Science as part of an Animal Technology programme.

Where a more structured approach has been adopted by employers, two general systems have been used. The first, appropriate to colleges and universities, consisted of a series of lectures and practical lessons during the summer vacation. The second consisted of transferring the trainee through several departments or sections to provide a range of skills in different disciplines.

The first of these methods, the 'vacational' training, is of limited value as the training conditions are unrealistic and training takes place not when there is a need but to suit the academic year and possibly to give employment to staff with little else to do during the summer.

The second approach of transferring the trainee through several departments or disciplines offers the advantage of providing a broad range of skills under practical conditions. However, much will depend upon the willingness of the section supervisors to provide an effective training experience for the young people rather than merely using them as unskilled pairs of hands. It also requires a degree of co-ordination of the training to ensure that individual supervisors do not conflict in the methods or subject matter taught.

Such schemes may be most useful at a prevocational stage as they allow the young person and the employer to decide whether a career as a technician is appropriate and also indicate the disciplines or areas most suitable for employment for each individual.

For these reasons the system of transferring a young person through several unrelated departments is probably most suited to those employed as supernumerary staff, *i.e.* not held against a particular established post. Once the basic training has been completed the young person can then be transferred to a suitable department to specialize in particular skills.

This broad skills approach is also most useful in respect of the Youth Training Scheme (YTS) and this will be considered in more detail later.

## TRAINING PROGRAMMES

Within the public sector and much of industry, technical staff are required to complete three or more years in the training grade, as trainees or juniors, and obtain a formal qualification before progressing to the lower ranges of the qualified structure. The system is based on time served rather than on objectives achieved.

If the training period is to be effective the supervisors, in consultation with specialists, will need to prepare a training programme and monitor progress. The first essentials are a training syllabus, an agreed programme and a log or record book in which to record the trainees' progress.

A number of approaches have been adopted by organizations faced with this task. The Wellcome Foundation is among those that have prepared detailed Laboratory Manuals designed to provide a nucleus of information and experiments related to the work of the relevant laboratories for each trainee. This is provided in loose-leaf format capable of expansion by the trainees who are expected to add notes and details of methods as they progress through the programme. The manual provides a step by step guide to the use of equipment and techniques as is shown by Figure 9.1.

At the other extreme there is the Simple Syllabus Tick Sheet system. The supervisor using this method merely ticks off completed sections of the training programme on a prepared syllabus (Figures 9.2 and 9.3). While these systems have the advantage of reducing the time and

## The Top-Pan Balance

This type of balance is very popular in laboratories, being very robust and easy to operate. They are made in a range of sensitivities and capacities. The one in the Training Centre is a Mettler which has a total capacity of 1,300g and is sensitive to 0.01g.

Directions for Weighing
NB   DO NOT LEAN ON THE BENCH—THIS WILL UPSET THE BALANCE

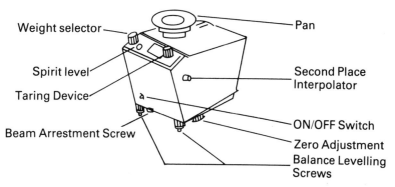

Step 1:   Make sure that the beam arrestment screw is turned fully to the left.
Step 2:   Plug into 13 amp socket.
Step 3:   Ensure that the bubble is central in the spirit level device. If YES go to step 5; if NO go to step 4.
Step 4:   Adjust one or both balance levelling screws to bring bubble to centre. *NB* This may require a bit of patience.
Step 5:   Depress ON/OFF switch—illuminated scale should appear in window.
Step 6:   Adjust taring knob fully anticlockwise.
Step 7:   Turn second place interpolator knob fully away from you.
Step 8:   Turn weight selector knob to zero (0).

**Figure 9.1** *Detailed instruction sheet.*
Reproduced by permission of the Wellcome Foundation.

effort required by the supervisor in monitoring the trainee's progress to the minimum it offers no information in respect of standards achieved, type of equipment used or precise procedures.

The system may be improved by considering the trainee's bench record book as part of the official record but after several months' use at the bench such books are often in far from pristine condition and may be a positive safety hazard!

| Haematology | Date | Supervisor |
|---|---|---|
| All standard haematological techniques including:<br><br>Blood counts (red, white, differential) using manual and automatic methods | | |

**Figure 9.2** *Simple syllabus training record book used by a university.*

| | Tick column when completed | Supervisor |
|---|---|---|
| Weighing | | |
| Balances | | |
| method of use<br>cleanliness<br>levelling<br>zero set direct reading scales<br>riders<br>use of forceps<br>errors<br>weights<br>weighing vessels<br>scoops | | |
| Types | | |
| semi-automatic<br>automatic<br>chemical | | |
| Special techniques | | |
| hygroscopic and efflorescent<br>    substances<br>liquid<br>*etc.* | | |

**Figure 9.3** *A more detailed tick sheet record book.*

An improvement on this system is provided by the use of a syllabus booklet combining a simple syllabus with a space to record a reference to a separate trainee log-book in which the trainee's achievements are written (Figure 9.4). The disadvantage of this is that neither book is meaningful on its own and the syllabus sheet could be considerably simplified as it in fact only acts as an index of skills obtained.

Photographic Techniques

| | Reference to log book demonstrating competence | Date and Supervisors signature |
|---|---|---|
| Safe methods of working | | |
| Preparations of graphic material for publication | | |
| Maintenance of equipment | | |
| Black and white photography including developing, processing, printing and enlarging. | | |

**Figure 9.4** *A syllabus booklet with space for recording reference to the trainees log-book.*

A number of training boards have extended the system of separate syllabus and log-book to include brief details of related theoretical knowledge with the practical training syllabus. This offers the advantage of producing a training syllabus, or recommendation, which will stand on its own as a useful publication and will offer the supervisor guidance when providing the training. Figure 9.5 shows an extract from the publication '18 Print Finishing' and Figure 9.6 a portion of the Chemical and Allied Products Industrial Training Board Recommendation for Physical Measurement Techniques.

The 'Institute of Science Technology (IST) Training Manual for

| Skills | Knowledge | Remarks |
|---|---|---|
| (1)  Assess prints for retouching | | |
| Identify and use correctly the company's standard lighting under conditions which work is to be viewed | Effects of viewing under non-standard conditions *e.g.* false impression of density or colour, colour casts from walls or surrounds | |
| Interpret job requirement from instructions | Layout of instruction sheets and information given with work *e.g.* (a)  delivery time, (b)  finish required (c)  mounting instructions (d)  special retouching requirements and the extra time allows. | |

**Figure 9.5** *Part of the photographic training booklet showing the identification of skills and related knowledge.*

| Practice/Skill | Related Knowledge |
|---|---|
| Sampling Obtaining a representative sample. | Techniques used. |
| pH Use of meters, papers, comparators, *etc.* | Meaning of pH and measurement. Use and care of apparatus and electrodes; special precautions. Buffer solutions. Sources of error, *e.g.*, foreign matter in sample. |

**Figure 9.6** *Physical measurement techniques.*

Technicians' (Weston, Heinemann Medical Books Ltd.) combines a number of systems by providing a simple syllabus checklist in the form of a 'programme planner' with a detailed training programme and syllabus recording sheet. The programme planner (Figure 9.7), offers a means of selecting an appropriate training course for each individual trainee and provides a quick method of quantitatively recording progress. The assessment record pages give a more detailed training syllabus than is used in many systems while allowing a whole page

BASIC SAFETY
1 Safety organization
2 Hygiene and safety
3 Accidents and first aid
4 Chemicals
5 Equipment

GENERAL LABORATORY
PROCEDURES
1 Apparatus cleaning and use
2 Calculations
3 Chemical storage
4 Connections
5 Cryogenic procedures (basic)
6 Filtering
7 Gas burners
8 Gas cylinders
9 General housekeeping
10 Holding equipment/supports
11 Laboratory use and storage
12 Measurement of volume
13 Records and recording
14 Services, electricity
15 Services, gas
16 Services, water
17 Services, vacuum
18 Solutions
19 Stores and stock control
20 Temperature reading/recording
21 Washing glassware, *etc.*
22 Work planning and costing
23 Audio visual
24 Reprographic processes
25 Classes and demonstration

BASIC LABORATORY
EQUIPMENT
1 Autoclave
2 Balances and weighing
3 Centrifuges
4 Chart recorders
5 Homogenizers and cell
disintegrators
6 Fume cupboards
7 Heating equipment
8 Light microscope
9 pH meter and measurement
10 Pumps
11 Ovens, incubators, *etc.*
12 Refrigerators and deepfreezes
13 Coolers/cold baths
14 Stirrers and shakers
15 Stills, deionisers, reverse osmosis
16 Water baths

RADIOCHEMICALS
1 Techniques and basic safety
2 Storage and labs
3 Use and disposal

INSTRUMENTATION
1 Amino acid analysers
2 Chloride analyser
3 Chromatography
4 Colorimeters
5 Computers (use of)
6 Computing
7 Electrophoresis
8 Electrophoresis/chromatogram
scanner
9 Fluorimeter
10 Ion exchange meters
11 Nephelometer
12 Osmometer
13 Oxygen meter
14 Particle counter
15 Polarimeter
16 Polarograph
17 Photometric analysis
18 Refractometer
19 Spectrometer
20 Stroboscope
21 Thermal analysis
22 Ultrasonic equipment
23 Viscometer
24 X-ray diffraction

BIOLOGICAL TECHNIQUES
1 Blood and body fluids
2 Centrifugation of blood

BIOCHEMISTRY
3 Biochemical assays
4 Commercial test kits
5 Cell counts
6 Cell harvesting
7 Manometric technique

PHYSIOLOGY
8 Isolated tissue preparations
9 Gas analysis
10 Organ baths
11 Oscilloscopes, *etc.*
12 Thermisters
13 Haematology
14 Blood Grouping

**Figure 9.7** *Part of the programme planner and record of training.*
Reproduced from 'Technician Training Manual', Heinemann Medical Books.

width for the trainee to record the precise details of the techniques used (Figure 9.8). The trainee's recording provides a qualitative method of assessment when taken with the performance level noted by the supervisor. The IST system also provides for the recording of levels of attainment, based on terminology used by CAPITB, using three skill levels of Appreciation, Practitioner and Expert. Of greater importance to the supervisor is that the use of this system keeps administration and monitoring procedures to the minimum.

## THE TRAINING PROGRAMME CONTENT

It is generally accepted that the training programme for school leavers should consist of three main sections or elements, using the terminology of the IST Manual; induction, basic technical training and further specific training related to a particular laboratory or job. Even where this concept is not accepted analysis of the training given will usually identify the components of these three elements. The 'IST Training Recommendations for Educational and Research Technicians' which complements the manual mentioned above, makes the suggestion that training programmes for laboratory staff should include, in addition to the relevant specialized skills, at least seventy percent of the items listed under the headings of Safety, General Laboratory Procedures and Basic Laboratory Equipment sections of the manual. This view is supported by surveys of supervisory technicians conducted by Weston in 1981/2 ('Istox Laboratory Management', booklet 6).

The point is made that while there may be a case for restricting the training of certain groups of technicians to their own specialization *e.g.* animal husbandry, photography and engineering, such narrow training may limit the career opportunities available to these staff at a later date. This may cause difficulties for both the employer and employee should redeployment prove necessary in the future.

It is therefore recommended that for school-leavers supervisors should adopt a broad skills approach to the basic training of technical staff in order to provide a firm grounding for the specific skills that will be required during the trainee's career.

Supervisors must also accept that not every unit or section is suitable for a training placement for a variety of reasons. These include:

• *Techniques.* The range of techniques used may be too restricted or of an inadequate standard *e.g.* maintenance workshop rather than true

| | Brief details of procedures, equipment, chemicals *etc.* | Skill level | Trainer | Date |
|---|---|---|---|---|
| **MEASUREMENT OF VOLUME**<br>Selection of suitable apparatus for use, compatibility of items chosen, *i.e.* it is not consistent to add 1 cm$^3$ accurately from a pipette to 2 litres measured in a measuring cylinder. | | | | |
| **Measuring cylinders, graduated flasks, and beakers**<br>Pouring and mixing procedures for different chemicals and solutions of acids and water. | | | | |
| **Volumetric flasks**<br>Principle and accuracy, graduation and miniscus. Marking on flask.<br>Use of flask, correct fit of stopper, cleanliness.<br>Heat and damage and drying after use. | | | | |
| **Burettes**<br>Care and use of burette, glass and PTFE stopcocks, retaining same in position during use.<br>Burette support systems, correct clamping.<br>Experience filling.<br>Storing of burettes. | | | | |
| **Pipettes**<br>Pasteur—<br>   Safe storage—danger from broken ends<br>   Use of plugged and non-plugged<br>   Holding and dispensing technique<br>   Disposal<br>   Use and advantages of plastic pipettes.<br>Graduated—<br>   Sizes available—graduations 'contain' and 'deliver' uses.<br>Volumetric—<br>   Recognition and use in measuring precise volumes. | | | | |
| **Pipette teats, bulbs and controllers**<br>Importance of safe pipetting techniques.<br>Types of bulbs and controllers available, suitability for particular pipettes and uses.<br>Experience in dispensing using a range of pipettes and controllers. | | | | |

Approval of Dept. Rep.        Training Officer

**Figure 9.8** *A sample page from the IST Training Manual.*

mechanical engineering providing a series of odd-jobs rather than experience in basic skills.

● *Facilities*. The actual physical facilities of the laboratory may not be adequate. The Medical Laboratory Technician Board in their leaflet 'Laboratories and Laboratory Based Training' (The Council for Professions Supplementary to Medicine, 1979) states that the Board has 'the right to refuse approval to any institution which is not properly organized and equipped'. In respect of general standards of laboratories they require at least 3 metres of bench for each member of staff with 4.5 square metres of unobstructed circulatory space.

● *Workload*. The workload in the section may be too great to allow time for a young person to be adequately trained in basic skills. The need may be for additional skills training for an existing member of staff or for the recruitment of another trained person rather than a young trainee.

● *Lack of Training Expertise*. The atmosphere within the section may be wrong possibly due to lack of management skills on the part of the supervisor. The position may be further complicated by the section supervisor being unaware of their own deficiencies. Indeed, a situation has occurred where a succession of young people placed as trainees in a section have left before completing their programme, without an appreciation that the supervisor was the cause of the trainee's 'failures' or inability to fit into the section.

Thus success of any training programme depends ultimately on the quality, competence and enthusiasm of the trainers or supervisors. It is important for the instructor to appreciate that training is basically a 'hands on' exercise and that the trainee should obtain practical experience in all aspects of the techniques involved, including the preparation of materials, care of apparatus, clearing away after use, basic maintenance of equipment and fault detection. Although it may be necessary for a technique to be taught in isolation *e.g.* use of centrifuge, importance should be placed on the ability to perform the task as part of a normal work sequence.

## Completion of Training

The formal training period for such trainees is normally three years and is usually linked with the need to acquire an appropriate BTec or Scotvec qualification. A number of employers are now moving away from the 'time-served' approach to training in favour of a 'skills acquired' system.

Some employers award a certificate or record of satisfactory training and the Institute of Science Technology (IST) Educational and Research Sector scheme offers a certificate endorsed by both the Institute and all the employers participating. The award of such a certificate provides a symbol of recognition of the trainee's achievement and as such provides a means of motivating staff.


## YTS IN LABORATORIES

The Manpower Services Commission Youth Training Scheme was introduced nationally in September 1983 to provide 16 and 17 year old school leavers with a one year programme of planned work experience integrated with work related or further education. This was extended to become a two year programme in 1986.

Each scheme includes:

(i) a period of induction and assessment of the trainees' needs,
(ii) planned work experience,
(iii) off-the-job training,
(iv) occupationally based training,
(v) basic skills of employment: the YTS core element of communication, number and its application, the use of equipment, problem solving and planning and introduction to computing and new technology.

The scheme is financed by the MSC and offers employers the opportunity to help young people while benefiting to some extent from the work they do during the work experience element. Figure 9.9 shows the simplified YTS structure. Recent changes have seen the MSC becoming the Training Commission and modifications of the structure, but the role of the supervisor has remained the same.

The technician supervisor may be involved in a number of aspects of YTS, depending on whether the supervisor arranges the whole scheme, or merely provides placements for a scheme operated by a local college or agency. In the latter case the supervisor's involvement may consist of ensuring the trainee receives, and records in the log-book, adequate work experience and liaison with the college. In the former situation the supervisor may be required to prepare the training programme, arrange for the provision of the off-the-job element (either within the organization or at a local college), submit the scheme for the approval of the Area Manpower Board, and ensure its smooth operation.

Such supervisors will be required to ensure that the scheme meets the following quality control criteria:

- *Occupationally Based Training.* The written programme should cover a broad but related range of work areas, it must indicate where the training will take place, *i.e.* on or off the job, the objectives, and who will be responsible for the training. The responsible persons may be the section supervisors or a subordinate technician.

- *Planned Work Experience.* Details as to the learning opportunities, integration of work experience with off-the-job training/education, the means by which trainees will be informed of the content, and monitoring procedures should be included.

- *Off-the-job Training or Education.* It will be necessary to show how this element has been integrated as part of the whole scheme and the content attendance and liaison arrangements demonstrated with the providers if this is a college or outside body. Within the technical sphere the educational element would normally consist of attendance on BTec or Scotvec courses and as such should not present difficulties.

- *Core Areas.* There is a need to show how, within the off-the-job and work related situations, provision will be made for the trainee to make progress within the core area topics.

- *Induction.* The induction should be planned to ensure that the trainees understand YTS, their role within it, the financial aspects of the scheme together with details of safety rules and conditions.

- *Assessment Review, Recording and Support.* The requirements for the supervisor in respect of these aspects of YTS are similar to those required by employed trainees. There will be a need to show the means of assessment, review and recording of progress within the organizational structure to ensure the scheme functions effectively, including counselling of trainees.

Upon completion of the YTS programme the trainee should be issued with a leaving certificate which must include details of the occupational skills and knowledge acquired, the accomplishment in the core areas and any other activities in which the training has been involved.

## WORK EXPERIENCE AND TRAINING ON EMPLOYERS PREMISES

There are a number of occasions in addition to YTS where supervisors may be faced with the responsibility of providing learning opportunities for young people who are not employees. Work-based learning is seen to offer young people a number of advantages. It provides a relevance to the subject which may not be fully appreciated at school, creating a need to know. In addition it offers motivation, as the young person is part of a team in a real situation and this is seen as an important factor for those who fail to realize their full potential in the classroom. However, such placements, which usually provide a mix of full-time education and employment experience, do provide the supervisor with additional problems. The short periods in which the young person is available at work means that if they are to be provided with an opportunity to learn 'by doing', rather than merely observing, they will require considerable attention from existing staff. Work schedules are seldom changed to take this factor into account with the result that the young person is given routine tasks to keep them occupied rather than to provide learning experiences. Under these circumstances it will be a measure of the supervisor's skill in motivating existing staff to undertake the considerable additional work involved in offering such schemes. Failure to explain the functions and motives of these schemes have often, in the past, resulted in them being seen to involve 'cheap labour'.

Before agreeing to act as host for such people the supervisor should decide on suitable tasks designed to provide a useful learning experience. This should be related to the needs of the group, offer immediate feedback and be of sufficient variety to stimulate interest. The Tavistock Guide 'Learning at Work' (MSC, 1982) discusses these problems at length and is recommended to those supervisors participating in TOPS, the Certificate of Prevocational Education and other work experience schemes.

It is essential to the success of such programmes that the supervisor establishes a direct link with the college tutor responsible for the educational element. All too often contact between college and employers occurs at administrative or head of department level with little or no direct contact between those engaged at the sharp end. If necessary the supervisor should by-pass the formal procedures and establish informal contacts.

**THE SAFETY OF TRAINEES**

The role of the supervisor in ensuring the health and safety of staff will be considered elsewhere. However, school leavers, because of their youth and inexperience, present particular safety problems. This is particularly the case of YTS trainees who may be placed in sections where the trainers may not be experienced in supervising young people. It is important that the trainers are not only aware of their responsibilities but that they should not minimize the risks of accidents.

The MSC gave eight points of general advice for supervisors of trainees. These are presented below.

(1) *Previous Knowledge.* Do not assume that the trainee knows how to do a particular job even if they have covered similar skills at college or in previous placements. Each laboratory will present additional hazards and slightly different procedures.

(2) *Understanding.* Ensure trainees fully understand what they are to do and how to do it before they start a task. Check their understanding by means of open questions, avoid those that require only yes or no answers.

(3) *Emergencies.* Explain the emergency procedures before the trainee has cause to discover that he does not know them. In addition to emergency exits and fire procedures this should include use of 'spill kits' when chemicals or biological hazards are encountered, first aid/eye wash equipment and electrical isolating switches.

(4) *Logical Approach.* Encourage the use of a mental checklist approach to the safety of machinery, equipment and procedures. This will avoid overlooking those items that have become second nature to the experienced technician *e.g.* noting position of guards, checking glassware for cracks.

(5) *Timing.* Allow time. The trainee will be much slower than the experienced technician and may need to repeat certain steps.

(6) *Safe Systems of Work.* Always work safely. Experienced staff often take short cuts and may ignore safe working practices particularly the use of guards and safety equipment.

(7) *Supervision.* Ensure that the trainees are not put at risk by working unsupervised, even for short periods, or by undertaking tasks that are too complicated for them.

(8) *Review*. When the trainees do become competent at a particular task the trainer should notify the supervisor and perhaps more importantly, problems or difficulties should be reported immediately they arise.

Although YTS and the Certificate in Prevocational Education are still relatively new concepts they are in fact fully compatible with existing systems of training within laboratories and may be adopted as an extension, or in practice as an introduction to, the existing training schemes.

The introduction of these schemes could provide the opportunity to critically examine and reassess attitudes to training, based not on tradition (or habit), but on training needs. Such scientific or analytical approaches to training (Figure 9.9) are not new, they just have not been used regularly in scientific and analytical laboratories.

## ANALYTICAL APPROACH TO TRAINING

In many organizations employing technicians, there appears to be little attention paid to making the supervisor aware of the overall manpower policy of the organization except where that policy is not to recruit or replace staff. In these circumstances the supervisor is made only too aware of the consequences of the policy. In the normal situation it would appear that provided the head of section has sufficient influence, and funds, staff are appointed almost automatically to fill vacancies or to enlarge the 'empire' with little consideration to the amount of work to be done or for the needs of other, less influential, sections. Under such circumstances it is the usual practice for replacement staff to be appointed to the same grade as their predecessor or held as a trainee against that grade, or in the case of new posts, for the appointment to be made against a job description written down to a grade in job evaluated structures rather than on the basis of actual need.

The use of such structures and in particular bench mark job descriptions as in the 'Blue Book' (Manual of Implementation, 1972) encourages an unthinking approach to technical staff job descriptions as they provide a simple pattern into which jobs can be filled by juggling words on a piece of paper. The alternative of careful job analysis, if considered at all, is seen as entailing too much work or as the cause of administrative inconvenience in changing the job grade (and consequently budget), but it is the key to developing meaningful jobs and a

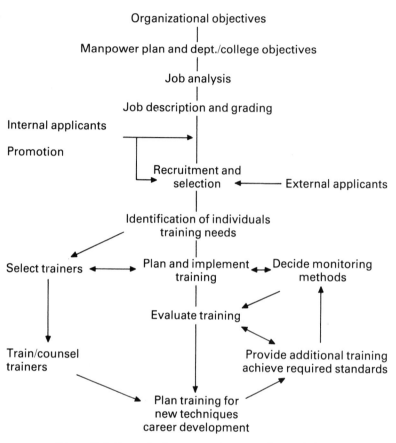

**Figure 9.9** *Stages in the systematic approach to training.*

sensible training policy, for training needs can only be assessed if the job description is accurate, realistic and complete.

### Job Analysis

Job evaluation techniques are known, if not fully understood, by many technical supervisors and they will have had experience in job descriptions as required by their employing body. Job analysis is the process that should be undertaken before the job description is written and can be considered as the total study of a particular occupation during which the relevant factors for each job are noted. These consist not only of the tasks and procedures undertaken by the job holder but the responsibilities and personal qualities that are required.

In the case of a new or vacant post, it is useful at this stage, to ask a number of people in the section what they consider the required tasks will be. It is often the case that colleagues have a better knowledge of the actual detailed duties than management or senior supervisors who would normally prepare the job description in those organizations where specialists are not employed to perform this task. The job description should be renewed every time a vacancy occurs and preferably during the annual counselling sessions, if held, or as changes take place, *e.g.* the appointment or retirement of scientists.

**The Job Description**

The job description (Figure 9.10) normally consists of several parts which will be considered separately.

*Title, Grade and Department.* These are for identification purposes and may also include the number of people performing the job. It is advisable to be consistent in the use of titles as these affect the status of the job holder. Where a job-title and grade identification are given *i.e.* Technician Supervisor grade 6, the use of the title when referring to the individual is more likely to have a positive effect than merely the depersonalizing grade number, unless the senior members of staff also abandon the use of titles.

*Function.* A short statement of the main function or objectives of the job. There are many instances where the job title or grade identification will give no indication of the scope or purpose of the job. This section should be kept short and should not include the duties or tasks which follow under different headings.

The function of technician posts could be to provide:

(i) a technical service to a research unit and maintain the laboratories as required, and
(ii) provide a photographic service to academic members of the department and audio-visual support in the main lecture theatre.

This information may be used to enable similar jobs with different job titles to be quickly identified *e.g.* within the same organization the storeman/head porter may be known as the Laboratory Steward while in other departments the Laboratory Steward may be the Chief

JOB GROUP 3: Electrical engineering                              JOB NO: 3/2

TYPE OF JOB: Electrical workshop/laboratory technician II        Grade: 5

Responsible to: Workshop Supervisor. Grade 7

Responsible for: No direct supervisory responsibility

MAJOR ACTIVITIES AND RESPONSIBILITIES

(1) General security of equipment, apparatus, and of laboratory areas.

(2) Installation and maintenance of electrical equipment, machines, consoles, pumps *etc.*, of research/teaching equipment for departmental use.

(3) Fault location and rectification of electrical equipment in workshop/laboratories.

(4) Construction, modification, development or servicing of apparatus for teaching and project work, using the full range of workshop techniques.

(5) Laying out of experiments for all levels of teaching work.

(6) General advice and assistance to students and assistance with tests and experiments.

(7) Some training responsibility as directed in respect of junior technical staff and occasional delegated supervisory responsibility.


NORMAL EDUCATION REQUIRED &
FORMAL TRAINING ACCEPTABLE:        BTec Higher National Certificate


NORMAL BACKGROUND EXPERIENCE REQUIRED
(before attaining lowest level of
acceptable job performance,                              7–9 years
including training period)

**Figure 9.10** *Simple job description for an electrical workshop technician.*


Technician, Departmental Superintendent or Manager. The 'function' on a job description also provides basic information to applicants for a post as to the scope of that post.

There is also a case for including a number of items not shown in the 'blue book' benchmark job description such as:

*Hours of Work.* Where training is organized by a Training Officer who may not be familiar with the hours of every grade of staff or the requirements of each unit this section will be of assistance in allowing training programmes to be prepared. As with the job function the inclusion of hours of work on the job description will be of value to applicants.

*Responsibilities and Authority.* Perhaps one of the major problems for supervisors in the public sector arises from having too great a responsibility and too little authority. The responsibilities refer to those members of staff for whom the person is responsible and it would normally be expected that they would have authority over the same people, being accountable for their duties. It is however often the case that technical supervisors are said to be responsible for a service or a section but do not have authority over the people providing or manning that service.

The second aspect of the responsibility part of the form relates to the person to whom the individual is responsible. These three sections as well as assisting the Training Officer to understand the structure in which the individual works, have the additional and most important function of making the structure clear to the individual concerned. This latter aspect is often confused in some laboratories particularly where job descriptions and organizational structures are considered to be confidential.

The main part of the job description consists of a list of duties and responsibilities (Figure 9.11). In some organizations, Duties and Responsibilities are given separately, but such duplication is often unnecessary. There is a strong case for the duties/responsibilities section including only those items undertaken by the individual concerned and not by subordinates, with the work done by others included under the heading 'Responsible For'.

The duties and responsibilities are normally listed on the job description in order of magnitude or frequency *i.e.* the minor or infrequent duties listed last. In some job descriptions this distinction is formalized with separate headings for Routine/Regular duties and Occasional/Irregular duties.

It is also possible to include the times during which certain duties must be carried out if the job is of a routine nature and there is a need to avoid the individual adopting a flexible approach.

**The Job Specification**

The job specification details the skills and knowledge required by the technician to perform each task. Normally this is prepared in conjunction with, and is based upon, the job description but there is no reason why it cannot be prepared separately for specific tasks. Indeed, where the only formal training programme within the organization is that given to young trainees who are moved through different sections the supervisor will need to prepare specifications for each task the trainee will undertake.

The approach has been adopted in the past by a number of Training Boards which have adopted the procedure of separating tasks, skills and knowledge in their training recommendations (Figure 9.11).

| Task | Skill | Related Knowledge |
|---|---|---|
| Care and Maintenance of hand tools | | |
| (1)  Recognize the need to resharpen | Identification of need for resharpening, damaged edges. | Types of tools Correct forms and angles. |
| (2)  Sharpening tools (a)  Using oil stone | Use of oil stone to sharpen, recognition of correct form. | Types stone, oil Correct stance |
| continued . . . | | |

**Figure 9.11** *Skill specification.*

In many technical job specifications this format will be deemed adequate, in others there will be a need for social skills. The emphasis placed on such skills will vary; normally the higher the job the greater the need for expertise in social skills when dealing with subordinates. A number of the lower grade jobs will also require such skills, where for example a technician provides a service to research groups or receives samples from clients.

The job specification should also include items necessary for the correct performance of the job. These may be physical requirements such as sufficient strength to lift workshop materials, absence of colour blindness in microscopists, absence of allergy to fur in animal technicians, or they may relate to certain skills which if not processed by the individual will indicate a training need.

Other factors may relate to length of experience and qualifications required for the job. In many cases these items will be covered by agreement between management and trades union but a certain amount of flexibility may be given to the supervisor.

The systematic approach to training has been considered in some detail as the technical grading system is frequently based upon a job evaluated structure, implying the existence of job analysis which, if used correctly, enables the identification of skill and knowledge requirements for specific posts.

There are a number of other possible approaches which will be recognized by the supervisor. More complicated, or structured approaches may be used including the use of questionnaires, the keeping of diaries by trainees, confrontation meetings and role set analysis. These are normally not suitable for use by the technical supervisor and are as such beyond the scope of this book.

**Administrative Approach to Training**

Many laboratories, particularly those without a separate Training Section, within the Personnel Department, adopt a purely administrative approach to training. Staff are expected to follow existing precedent and work within the formal system. Emphasis is placed on formal courses, usually of technical education rather than training, with little or no identification of training needs, diagnosis of weakness, or evaluation of results. Usually the administrators arranging the college attendance will have no knowledge of the actual syllabus content and may be unaware of changes in curriculum or examining body. This approach may also be classified under the title of 'The passive approach' or 'Training by chance'.

**Welfare Approach**

A more enlightened view is shown by organizations that adopt what may be described as the welfare approach. In these organizations all training is considered to be of value and courses are approved for all who wish to obtain a recognized qualification. Usually this results in all recruits to the organization being expected to follow a training programme or a course of study. The system does tend to offer technicians a well-planned career structure with emphasis being placed on promotion on the basis of examination results, with the possibility of insufficient attention being placed on practical skills. Under this approach there is often confusion between the education and training elements with the result that training in practical skills is often neglected.

The welfare approach could offer the technical supervisor the chance to influence junior technical staff and local colleges in the choice of appropriate courses.

The principal disadvantages of this approach is, as for the administrative oriented system, that the 'training' provided is not directed to achieving the organizational objectives and may not even be related to them. However, it does offer a degree of satisfaction to the technical staff, provided the supervisor ensures that the courses they attend will meet the career needs of the staff.

## LEVELS OF TRAINING NEEDS

Boydell in 'The Identification of Training Needs' (BACIE, 1979) identified three levels of training needs.

### The Organizational Level

This relates to general weaknesses which indicate where within the organization training is needed most. In many respects detecting such needs, in areas outside the supervisory technician's control, is easy. Rectifying the situation is more difficult as the supervisor will lack the authority to take action. Where Training Officers are employed, it should be possible to draw the problem to their attention, but within many laboratories where the Training Officer is part of the administrative structure normally responsible for staff such comments may be taken as criticism.

The organizational training need relates to a weakness or shortfall that once identified can be rectified by training. Once identified at organizational level training may be required at the occupational level.

### Occupational Needs

Most supervisors will be familiar with thinking in terms of occupational training *i.e.* training in the skills and attitudes required by a particular job or occupation. Industrial Training Boards such as the Engineering ITB and the Chemical and Allied Products ITB (before it was disbanded) and professional bodies such as the Institute of Science

Technology have produced recommendations and manuals for new entrant training in occupational skills.

**Individual Needs**

Once the organizational and occupational needs have been determined it is possible to decide which individuals need training in what particular skills. This is done by identifying deficiencies in particular skills, knowledge or attitudes of individual members of staff. The question of attitude training may be new to many technical supervisors who may be used to relating training solely to practical skills. An illustration of this problem and the levels of training need is given in the following case histories.

• Case (1). There were a number of complaints from members of scientific staff using a central analytical service that the senior technician was off-hand and unhelpful. The training needs at the three levels are shown below.

(i) Organizational. The main area of need was the Analytical Section.

(ii) Occupational. Social and communication skills were required for specimen reception.

(iii) Individual. The senior technician needed training in social skills.

Further investigation may indicate that training in the submission of samples, timing of submissions, *etc.* were also required by the scientific users of the service but unfortunately this is beyond the scope of this book.

The problem indicated here, that of correctly identifying the real training need rather than just the cause of complaints, has been touched upon elsewhere. The difficulties are further illustrated by the example below.

• Case (2). The Chief Technician in an animal house provided new junior assistants with a short induction training but once this was completed expected them to carry out routine duties with no further training. Consequently many complaints were made by the scientific staff to the experienced technical staff (ONC level). This unsettled those staff, who then resigned, resulting in high staff turnover which in turn led to the Chief Technician having to assist with routine work, and due to pressure of work giving up the induction training. The younger staff then started to leave more frequently and the service decreased further. Eventually management became aware of the increasing level of complaints and sacked the Chief Technician.

It is felt that this example in which lack of skills training at a relatively basic level plus a deficiency in supervisory training resulted in a technical supervisor not doing his real job, holds a lesson for many technicians, particularly where the need for a supervisor to cover lower level work results from staff turnover or cut-backs.

## THE IDENTIFICATION OF PRESENT TRAINING NEEDS

There are symptoms relating to staff performance, recruitment problems and complaints from clients which may suggest a training need (Figure 9.12).

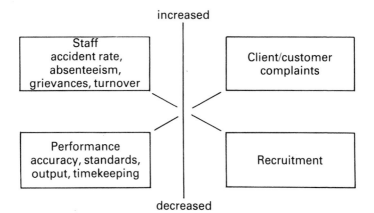

**Figure 9.12** *Symptoms of a present training need.*

While it may be possible to gain a general impression from these factors it is necessary to quantify the position before acting to improve the system. This may be done by comparison with past experience, other departments within the organization or the experience of colleagues. In all cases the supervisor may consult records which, although their primary function may not relate to training, may be used in quantifying one or more of the parameters under review, *e.g.* routine sterility test results are primarily designed to ensure that contaminated culture media is not used by the laboratory but they may provide a means of assessing the aseptic technique of the member of staff preparing the media.

Having identified the need, it is necessary to decide whether improvements can be made by means of training, and if so, whether the provision of that training is within the area of responsibility of the supervisor. The technical supervisor may often be aware of a problem but will lack the authority to deal with it. Within large scientific establishments the problems may be due to lack of organizational planning, lack of managerial skills or awareness on the part of senior scientific staff, absences of cost or inventory control systems, failures of communication or conflict between scientific, technical and administrative staff.

Organizational weaknesses frequently give rise to a large number of informal complaints and dissatisfaction among members of staff which do not reach the stage of a formal grievance. Detecting and acting upon such complaints is often difficult as the supervisor/subordinate relationship is part, and often a large part, of the problem. Such problems may relate to uneven distribution of the work load with the supervisor and his immediate group selecting the best jobs and delegating the less pleasant or the 'clearing up' to others. They may be based upon what is seen as an unfair system of promotion, again with the supervisor favouring his own group, or just a dissatisfaction with the organizational skills of the Chief Technician.

There are a number of methods by which such information may be collected.

## The Annual Review Interview

This is often the simplest for the supervisor to administer. Under this system the supervisor interviews each member of staff using a prepared programme to discuss their future, present progress, and feelings about the organization. As part of this interview the supervisor can include questions to identify training problems and needs. The disadvantage of this method is that the supervisor only gets a response to questions asked and may not identify other needs. The use of less structured interviews may overcome this problem but are more time consuming and require considerable interviewing skills on the part of the supervisor.

## Peer Interview

Peer interviews, in which small groups of staff are encouraged to interview each other, may be very effective as a means of identifying training needs particularly among more mature employees. Such interviews or discussions may be held at section level on a regular basis and could be held in conjunction with the Laboratory meeting at which work progress is reviewed.

## FUTURE TRAINING NEEDS

The supervisor should not confine the training to meeting present needs but should be active in seeking and identifying future training needs of the section. These needs, as the use of the word future implies, involve training to meet change be it change in objectives, product, technical processes, equipment, political policy, legislation or personnel. Of these the latter two are easier to recognize as producing training needs. The appointment of new staff particularly trainee technicians, produces an obvious need to train, while the Health and Safety at Work Act has resulted in a statutory duty to provide safety training. The supervisor should not however concentrate on these two factors to the exclusion of others.

### Changes in Objectives

Within any laboratory, objectives change together with factors which affect the ability of staff to achieve those objectives. The problem is that without some form of supervision by objectives technicians may not be able to recognize the change let alone the training needs necessary to meet the new challenge.

### Changes in Product

Product is defined in the widest possible sense in this context and not viewed solely as a manufactured product. The results produced by the laboratory may suffer a fall or increase in demand, or be overtaken by new technology requiring changes or abandonment.

### Technical Changes

Methods and processes will change as a result of technological improvements resulting in increased automation of techniques and complexity of equipment and an increased demand for a knowledge of electronics or computing among technical staff.

**Political Policy and the Financial Climate**

Staff cuts, reductions in expenditure and tighter budgeting as a result of the financial situation may require increased flexibility from staff if services are not to fall below acceptable standards. The political policy of the Government may also directly influence training as with the introduction of the Youth Training Scheme.

## SKILLS TEACHING

To be an effective trainer the supervisor will need some understanding as to how people learn skills, or instructional methods and the use of training aids.

### How People Learn Skills

The majority of people learn best by means of visual stimulation, by seeing rather than hearing. Learning is improved where more than one sense is involved. There is also an advantage in involving the trainee actively in the learning process rather than allowing him to be passive —encourage hands-on skills and practice.

It is often a good idea to change activities or allow a break when teaching. Most people learn quickly at first and then reach a learning plateau when being taught by traditional methods.

Feedback is important to the trainee and encourages learning, particularly when used in conjunction with 'reinforcement' by praise and recognition of the trainee's achievements. The objectives must be recognized as important by the trainee and the training seen as relevant. Recognition may be in the form of a useful skill, promotion, money *etc.*

Finally, the trainer must be supportive and should not criticize mistakes in subordinates. By all means point them out to the trainee but find something to praise as well.

Earlier in this chapter the traditional and analytical approaches to training were considered. These can be applied to the methods of skills teaching. The analytical method indicated below has the advantage of reducing the time taken to reach the experienced worker standard and reduces the learning plateau.

## ● The Traditional Method of Skills Training

Stage 1. The trainer demonstrates the whole task, and then repeats the demonstration with an explanation.

Stage 2. The trainee practises the whole task with the supervisor correcting mistakes. Practice continues until the performance is adequate.

Stage 3. The trainee is transferred to the work/production situation.

This method leads to the development of a learning plateau during which performance remains stable but short of the maximum level that could be attained (Figure 9.13).

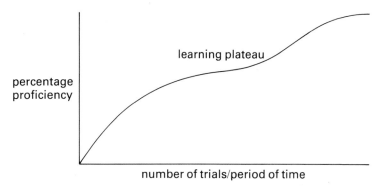

**Figure 9.13** *Learning curve—conventional training methods.*

## ● The Analytical Method

Stage 1. Supervisor analyses task.

Stage 2. Supervisor carries out skills analysis breaking task down into its components or elements (Figure 9.14).

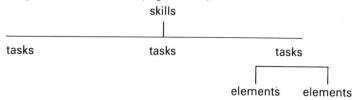

**Figure 9.14** *Skills analysis—the process of identifying and describing the sensory and motor elements of a task.*

Stage 3. Supervisor demonstrates and explains whole task to the trainee.

Stage 4. The supervisor then demonstrates the first skill element.

It will be seen from Figure 9.15 that there is an initial period (A) during which the skills are being learned where no productive work occurs. More importantly the learning plateau (area B) is overcome allowing satisfactory production levels to be reached quickly.

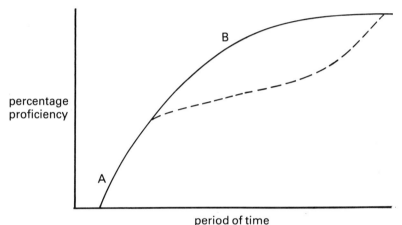

**Figure 9.15** *Learning curve—analytical training methods.*

Stage 5. Trainee practises the first element until proficient.

Stage 6. 4 and 5 are repeated with each skills element.

Stage 7. All the tasks are combined into the whole task and practised until overall performance is adequate.

Stage 8. The trainee is transferred to the production situation.

As will be seen, the analytical approach to skills training requires considerably more preparation and planning on the part of the supervisor in analysing the task and skills, but such preparation is seldom wasted.

**Training Materials**

Most technical supervisors will be experienced in training by practical demonstration and practice at the bench but will have relatively little experience in the use of teaching aids. In many cases such aids are considered the province of teachers and are held to be more appropriate to use as part of the day release or college element of the training programme. Even technicians involved in the use of 'hand-outs' and practical notes for students (at college or university) rarely prepare such material.

Information on the use of such aids as overhead projectors, slides and film can be found in any good textbook on teaching and those requiring further information are recommended to read 'Teaching in Further Education' (L. B. Curzon, Published by Cassell). The use of such aids is appropriate to the training of technicians in practical laboratory techniques and should form part of the systematic approach adopted by supervisors.

Video, which is increasingly being used in undergraduate and other teaching, offers a most effective tool to the supervisor who has access to the necessary equipment particularly as part of a programmed learning approach allowing the trainee to obtain skills at his own speed.

**Training—Follow-up**

Whatever approach is adopted towards training, some means of follow-up is essential to test the effectiveness of both teaching and learning. In the first instance the completed task needs to be examined and checked with the trainee, allowing time for questions and discussion.

As the trainee becomes more experienced the checks can be reduced and changed from direct examination to indirect methods. At this stage the actual training results should be compared with the original objectives. To ensure maximum information is obtained from this stage, the supervisor should discuss the system with the trainee and trainer separately. Finally modifications should be made as a result of the discussions and results.

**TRAINING OF SUPERVISORS**

At present most technical supervisors within the public sector learn the necessary supervisory skills by informal methods, *i.e.* practical experience and mistakes. Unfortunately this tends to perpetuate bad management as the inappropriate styles and methods of yesterday's senior staff are adopted by today's.

While the public sector has neglected supervisory training for technicians, industry has, for a number of years, been organizing short courses in specific skills for their own staff. Such programmes are either offered and organized on an in-house basis, or are held in conjunction with outside agencies, for example the MSC Training Within Industry Group (TWI), which also provided residential courses before it was abolished.

In addition a number of professional bodies such as the Institute of Supervisory Management offer examinations of a general nature, as does the Business and Technician Education Council. Of these examinations, organized specifically for technicians, perhaps the most widely used in the past was the BTEC Supervisory Studies units offered as part of the Laboratory Science and Administration programme. The disadvantage of this particular examination is that it is studied at HNC level by young people in their late teens or early twenties who lack the experience and maturity to obtain maximum benefit from the course.

The recently introduced BTEC Continuing Education programme in Laboratory Supervision and Management has been designed for more mature students who, it is hoped, will follow the course at a time in their careers when they need supervisory skills. Unfortunately this is still limited to one or two colleges. The authors have recognized the need for a more accessible course and, as part of Laboratory Supervisory Training Services (LSTS), have introduced a 12 unit distance learning programme in laboratory supervisory skills specially designed for technicians.

The Institute of Medical Laboratory Science and the College of Radiographers offers a specialist qualification in Management for their members, who are mainly employed in the National Health Service, as did the Institute of Science Technology although this examination has been discontinued in favour of the BTEC continuing education award and the LSTS course.

**Internal Courses**

The advantage of in-house supervisory courses is that they can be tailored to the precise needs of the organization, or at least to the needs as perceived by management; the two may not be the same. Such courses also offer a greater degree of flexibility, can be offered as part of career progression, and need not be governed by staff/student ratios. The lectures may be provided by experts from within the company, with specialist help provided from outside the organization in the form of guest speakers.

There is often a problem in convincing management that there is a need for supervisory training whether provided in-house or through examining bodies. However, the same is not true of senior technical staff. Barclay in 'Technical Managers Activity and Training Survey' (The Polytechnic Huddersfield, 1982) found that 'personal effectiveness training' was the largest need perceived by such staff, with additional needs for up-dating and problem analysis in technical areas. It is to be hoped that despite the present financial situation the means of satisfying these needs will be found.

# CHAPTER 10

# Counselling and Discipline

## INTRODUCTION

The need for counselling may arise from two types of problems, which while distinct in origins, may overlap in their effects upon the employee's work. The first are work related problems resulting in poor performance, unsatisfactory attitude and other matters which may, if allowed to continue, give rise to the need for disciplinary action. Such problems may arise from a number of causes real and imagined, including: lack of information relating to the future of the job, inability to perform certain tasks, over zealous or poor supervisor, feeling hard done by in comparison with colleagues.

The need for counselling may also arise from problems not related to work such as an unhappy marriage, illness, financial or housing problems. These may cause worries that prevent concentration at work and result in problems such as poor performance, and lack of attention to detail. Faced with such a situation the supervisor will need to ascertain the cause of the problem before taking any action such as issuing an informal warning, or other disciplinary measures, which might possibly exacerbate a personal difficulty.

The counselling interview offers a means of identifying the cause of a problem, of jointly recognizing its effect on work, and of deciding on a course of action to deal with it. Where the cause of the problem is identified as being 'personal' rather than work related, the supervisor should refer the individual to the person best able to help either within or outside of the organization. Within the company this may be the Personnel Department, Medical Officer, Pensions Section *etc.* or Citizens Advice Bureau, Guidance Counsellor or Family Doctor outside.

**The Counselling Interview**

The counselling interview should be similar in format to the disciplinary interview (discussed later in this chapter), except that it will require a more sensitive approach. This is often described as 'non-directive'. In practice the supervisor, as with other interviews, will need to direct the discussion but in a discreet manner.

The interview may be considered in five stages:

(1) *Introduction.* The supervisor introduces the problems that have arisen and if possible seeks agreement that they do in fact exist.

(2) *Identifying the Causes.* Attempting to identify the underlying courses of the problems by use of open questions and listening carefully to the interviewee.

(3) *Agreeing on action.* Where possible, agreement should be reached on an agreed course of action either to rectify the situation or to refer the problem for expert help.

(4) *Summarizing.* Prior to concluding, the supervisor should summarize what has been said about the identification of the problem and the agreed course of action.

(5) *Following-up.* Finally agreement should be reached on the date for a review of the situation to decide whether further action is necessary.

The supervisor should conduct the Counselling interview and subsequent follow-up with the intention of not only solving or helping with the particular problem but of creating a climate of confidence and care within the group.


**DISCIPLINE**

Discipline may be considered under two aspects; self discipline and imposed discipline. The latter will only become important when the former fails.

The supervisor should encourage staff to perform the tasks required of them by means of their own self discipline rather than by means of discipline in the form of controls and punishments imposed from above, and thereby develop professionalism amongst the workforce. Self-discipline coupled with good morale and high motivation are positive forces which encourage a positive attitude to work where attention will be given to accuracy in following procedures and interpreting results. Such standards will normally be higher than can be imposed by external forces.

There will be occasions, however, when it will be necessary to introduce external means of discipline. Such instances may arise from bad time keeping, specific occurrences of poor workmanship or as a result of appraisal. No matter how the situation arises it is essential that the action be fair, appropriate to the problem, and supported by those in higher authority should it not be settled at supervisory level.

Every supervisor should be aware of both the informal and formal disciplinary procedures of the organization and of the limits of the authority delegated to line supervisors. Imposed discipline may be necessary on occasions but it must not be forgotten that it will probably have a negative effect on the motivation of the recipient. This effect may be reduced by the manner in which the action is taken *i.e.* if it is felt to be fair and taken discreetly. Failure to take disciplinary action against an individual can have a negative effect on others who may feel that it is unfair if someone else 'gets away' with bad time keeping, poor work *etc*.

**Poor Time Keeping**

Poor time keeping is often associated with a lack of motivation and a poor standard of work. It frequently figures in disciplinary procedures, often only as a symptom rather than the main cause.

The situation is complicated within some educational and research sections where there exists a casual attitude towards time keeping, with considerable latitude being allowed provided 'the work gets done'. This attitude is often based on an understanding that technicians will work late to finish an experiment. It can be argued that this shows a flexible approach on the part of management and saves the organization overtime payments. In practice however it may simply be a case that the research scientist is himself a poor time keeper, spends mornings at college rather than the laboratory, or cannot reconcile his own contract to lecture to that of the technician employed to work specific hours.

Whatever the reason for the flexibility it can cause difficulties for the supervisor responsible for a whole department. Technicians working irregular hours may present a safety problem by working in the building when it is officially closed. Other staff employed in service or other sections where they are required to work their contractual hours may become disgruntled at the apparent unfairness of the system. The situation may also arise that any individual being disciplined for poor time keeping will feel that there are dual standards.

The technical supervisor should attempt to ensure that all staff work their contractual hours, if necessary having contracts changed to suit the actual situation. The actual monitoring of time keeping is best delegated to the first line supervisor with the chief technician making occasional checks in specific areas.

## Disciplinary Practice and Procedures

Disciplinary procedures are necessary for promoting fairness, to enable employees to know what standard of conduct is expected of them and in respect of the law relating to dismissal. Both the grounds for dismissal and the way in which it is handled can be challenged before an industrial tribunal.

Disciplinary agreements are normally negotiated between the employer and the relevant trade union, being based on the Advisory Conciliation and Arbitration Service (ACAS) Code No. 1. This code identifies a number of essential features of a disciplinary procedure in that they should:

(a) Be in writing.
(b) Specify to whom they apply.
(c) Provide for matters to be dealt with quickly.
(d) Indicate the disciplinary actions which may be taken.
(e) Specify the levels of management which have the authority to take the various forms of action.
(f) Provide for the individual to be informed about complaints and given an opportunity to state their case.
(g) Give the right for trade union representation.
(h) Ensure that, except for gross misconduct, no employee is dismissed for a first offence.
(i) Provide for investigation before action.
(j) Ensure that an explanation is given for any penalty imposed.
(k) Provide for a system of appeal.

## Informal Reprimands

The informal reprimand may be considered to be the stage in the normal supervisor/subordinate relationship that precedes the disciplinary procedure.

The informal reprimand is normally administered at the time of the 'offence' and is usually intended to be the end of the matter. Even at this stage the supervisor should attempt to make the reprimand acceptable to the person receiving it and not take the opportunity as a means of venting the supervisor's spleen or of demonstrating the dominance of their position. The importance of making the reprimand acceptable to the employee cannot be over emphasized. While the immediate aim is issuing the reprimand will be to point out a mistake or fault the objective must be to ensure an effective and well motivated work force *i.e.* to correct not punish.

Before issuing the reprimand the supervisor should establish the facts and background of the offence. If possible the supervisor should check the information and attempt to separate facts from opinions at this stage. The actual reprimand should be in private and controlled. The supervisor should tell the employee the reasons for the reprimand and then listen without interruption to the employee's comments. While it will not be possible for the supervisor, as the representative of authority, to be impartial they should be straightforward, fair, and avoid sarcasm and loss of temper. Where possible the supervisor should attempt to counsel rather than criticize. Departmental rules and regulations should be used as a guide but not followed slavishly. Each case should be decided on its merits. The interview should not be prolonged if the offender accepts the situation.

At the end of the reprimand the supervisor should attempt to be constructive, praising some aspects of the employee's work or discussing future developments. After the reprimand the recipient should:

(i) understand why the reprimand had to be given,
(ii) appreciate why the offence demanded a reprimand, and
(iii) feel that they have been treated fairly and in a similar manner to other people.

More serious offences, or where informal reprimands have proved ineffective, should be dealt with under the first stages of the Disciplinary Procedure.

## Disciplinary Procedure

The Advisory Conciliation and Arbitration Service (ACAS) have produced a Code of Practice relating to disciplinary practices in employment (ACAS Code No. 1, 1977) and this should form the basis of all local procedures.

*Informal Warning.* Informal warnings are normally administered by the immediate supervisor using the same procedure as for a reprimand but making it clear that the warning forms the first stage of the official disciplinary procedures. The supervisor should keep a note of the warning and any changes agreed. The Personnel Section should be notified of the warning, and support obtained should the next stage of the procedure be invoked.

The second stage of the procedure usually consists of an oral or written warning.

*Formal Warning.* The first formal warning may be issued for more serious, but still relatively minor offences, or following a failure to improve performance following reprimands and informal warnings. Although most disciplinary procedures are written to deal with serious offences the minority of cases will arise from persistent minor offences. The formal warning may be issued by the first line supervisor, by the manager or by a member of the Personnel Section. Where the procedure allows for both oral and written formal warnings as separate steps in the procedure, the oral warning is usually delivered by the immediate supervisor and the written warning by either the manager or personnel specialist if the problem continues.

Except in the event of an oral warning, details of the interview recording any disciplinary action should be given to the employee, with details of the time span in which an improvement must be made and any appeals procedure.

*Final Warning.* Further misconduct within the time limit given might warrant a final written warning, which should contain a statement that any recurrence would lead to suspension, with or without pay depending on the disciplinary agreement, or dismissal.

The supervisor should be present during each stage of the formal warning procedure, with the trade union representative present during formal stages.

At all stages of the disciplinary procedure the action taken must be able to satisfy the test of reasonableness and the supervisor must consider this before initiating the next step towards dismissal. Provided the employer has acted reasonably, it is often possible to reach agreement with the cooperation of the trade union so that an unsatisfactory employee can leave the organization rather than be dismissed.

In cases of gross misconduct, serious breaches of discipline or gross negligence it may be appropriate to summarily dismiss an employee. There should be a hearing at which the employee must have the opportunity to state their case in the presence of the trade union representative or a friend. Cases of summary dismissal should only be instigated by the head of department, or his deputy, and preferably only after consultation with the Personnel Officer.

In practice, even the most simple disciplinary problem can cause difficulties for the supervisor not only in respect to the actual disciplinary offence but by affecting motivation and working practices as the following example illustrates.

Animal technician 'A' was failing to perform his work adequately, to the extent that members of the scientific staff were asking that he should no longer look after their animals, and he was guilty of poor time keeping. His supervisor had spoken to him on both points without any change in behaviour and in response 'A' had by-passed his supervisor to complain to the Chief Technician that he was being picked on. The Chief Technician had no direct responsibilities for the animal house and took no action, but was always available to give a sympathetic response to any complaints.

Eventually the situation became so bad that the Personnel Officer, after repeated requests from the supervisor, agreed to see 'A'. Unfortunately, rather than follow the official procedure and provide counselling and/or an informal warning, it was decided to give 'A' a 'final talking to' in the presence of his supervisor and the Chief Technician.

The resultant interview was inadequately planned, with each of the 'external' parties having separate input to the extent that the supervisor found himself attempting to justify his complaints and working system before two arbitrators.

The final result was that 'A' felt that he was justified in continuing his behaviour pattern; the supervisor felt disgruntled at the lack of support provided by the department; and the Trades Union was in a position to complain if any further action was taken against 'A', such as the issue of a written warning, because the procedures laid down in the disciplinary agreement had not been followed.

## GRIEVANCE PROCEDURE

Grievance procedures may be used for disciplinary appeals but ACAS feel that it is more appropriate to keep the two kinds of procedure separate.

Normally the grievance procedure is used to deal with individual or collective complaints which may deal with gradings, difficulties relating to supervision, payment *etc.* The use of the word 'grievance' for what is in fact the appeals or complaints procedure, may discourage use of the system as individuals may not wish to be considered 'aggrieved' while they would be prepared to 'appeal'.

A large number of complaints made under grievance procedures will arise from the actions or inactions of supervisors which require the supervisor to be fully aware of the detailed procedures if the situation is not to be exacerbated.

Grievance procedures tend to follow the pattern of disciplinary procedures, passing through a series of stages which each involve a more senior member of management. It is normal for the aggrieved to be accompanied by a trades union representative at all stages.

The first stage may involve the grievance being raised with the immediate supervisor, although in some agreements the supervisor will be by-passed with the matter being taken directly to the manager or administrator. If the problem is unresolved at this stage the second level may involve a more senior manager, such as the head of department.

The third stage of the procedure may involve the central authorities, Personnel Officer or a joint management/trade union committee. There may be an external appeal system if the problem is still unresolved but this is relatively unusual.

In the case of 'collective grievance' involving more than one employee the first stage of the procedure may be by-passed with the problem being taken straight to senior management.

## TERMINATION OF EMPLOYMENT

### Termination by the Employee

Staff should be encouraged to inform their supervisor if they intend to seek fresh employment. This may happen as a matter of course if a reference is required. Prior to the employee deciding to leave, the supervisor may be able to offer help in obtaining advancement within the organization, in establishing contacts with future employers, or in rectifying problems. Help under these circumstances will not only assist the person who is planning to leave but is good for the morale of others.

In some organizations the Personnel Section will organize a termination interview to obtain the reasons for leaving. Over a period, examination of the reasons given can provide information indicating organizational areas that need to be improved *e.g.* promotion, prospects, job design. Where management do not arrange such interviews supervisors could organize them informally.

Reasons most frequently given for leaving are:

(i) promotional gain,

(ii) domestic problems,
(iii) dissatisfaction with the job or with supervision,
(iv) personal circumstances,
(v) better location.

When staff leave under these circumstances it is by mutual agreement between the employer and the employee, with the employee giving notice. The statutory minimum period of notice to be given by the employee who has been employed for four weeks or more, is one week although the contractual period of notice may be longer. In most cases, monthly paid staff will have a minimum contractual period of one month.

The employee may also terminate employment summarily if the employer is in breach of a fundamental term of the contract of employment. Such circumstances are described as 'constructive dismissal' for redundancy payments and unfair dismissal purposes. The employee being classed as having been dismissed within the meaning of the Employment Protection (Consolidation) Act.

**Termination by the Employer**

*Expiry of a fixed term contract.* If a contract is for a fixed period, as is frequently the case in universities and hospital research laboratories, it will automatically terminate at the end of that period. No notice is needed but the supervisor should approach such staff, preferably three months before the date of expiry, to discuss the options *e.g.* renewal, alternative employment within the organization, or help in seeking employment elsewhere.

If an employee remains in the post after expiry of the term of contract they will legally be considered to be working under the same terms and conditions except that the employment is subject to an implied term with an entitlement to reasonable notice.

The expiry of a fixed term contract may constitute dismissal in respect of redundancy payments and unfair dismissal legislation.

*Termination by Notice.* The contract of employment will usually specify the period of notice to be given to terminate the contract. Where there is a disciplinary agreement supervisors will need to ensure that the full procedure is followed before dismissal. The statutory minimum notice ranges from one week, for less than two years service, to twelve weeks for twelve years service.

*Summary termination for gross misconduct.* If an employee behaves in a way which is considered incompatible with the discharge of their duty they may be dismissed instantly without notice or wages in lieu of notice. Examples of such misconduct are disobedience of a lawful order, theft of employer's property, and drunkenness sufficient to impair the performance of their duty. The misconduct must be gross or grave in the particular circumstances of each case. As with termination by notice, the position will be clarified in the disciplinary procedure. Summary dismissal for conduct which is interpreted as neither gross nor grave will render the employer liable for damages for wrongful dismissal as a breach of contract.

*Illness.* In the past it has been accepted that the contract of employment will be frustrated where illness results in an employee being unable to perform as an essential part of it or where the illness lasts for a long time. It is difficult in circumstances such as an illness to decide whether 'frustration' has in fact occurred. An absence of 4 months may be sufficient (Hart and A. R. Marshall v Sons (Bulwell) Ltd. (1977) 1WLR 1067) in certain situations, while in others 18 months (Marshall v Harland and Wolff Ltd. (1972) 1CR 101) may not. If a supervisor feels that the work of the team is being seriously inconvenienced by the illness of a subordinate the matter should be raised with the Personnel Section. In practice, the opposite may occur with the manager or Personnel Officer wishing to take action against a subordinate while the supervisor is prepared to continue providing cover.

*Redundancy.* Unfortunately redundancy is an increasing problem throughout industry. The role of the technical supervisor in such matters is usually very limited, with the supervisor occasionally being involved in approving applications from subordinate staff for voluntary redundancy.

In situations where the employer is attempting to avoid the need for redundancy by restricting recruitment, the retirement of long service employees who are beyond the normal retiring age, short-time working, reduction in overtime, or transfer of staff to other work, the supervisor will need:

(i) to be involved in discussion with management on the implications of these factors on the functioning of the unit,
(ii) to be concerned with maintaining the motivation of staff, and
(iii) arrange for the retraining of staff transferred to the section.

Transferred staff will be under considerable strain resulting from the transfer and the prospect of learning new skills. The supervisor will need to be supportive in approach and make allowances for the stress to which the individual will have been exposed.

# CHAPTER 11

# Industrial Relations: the Supervisor and the Trades Unions

## THE STRUCTURE OF THE TRADES UNIONS

### Historical Background

The unions had their origin in the associations of workers which were formed to improve their pay and conditions. Such associations were originally held unlawful under Common Law and criminal under the Combination Act of 1800.

The Trades Disputes Act of 1906 gave the unions immunity from liability under civil but not criminal law. The Trades Dispute Act was repealed by the Industrial Relations Act 1971, which was itself repealed by the Trade Union and Labour Relations Act 1974. This act restores a general immunity upon Trade Unions from action in tort. The Employment Act 1982, in turn, removed this immunity and now, subject to limits on damages, trades unions are liable to the same extent as individuals.

### Types of Union

The present Law recognizes trades unions which consist of workers of one or more descriptions. The principal purposes include the regulation of relations between workers and employers or constituent organizations which fulfil those functions. This allows for the single union, *e.g.* National Union of Teachers, and the conglomerate made up of a number of sections such as the Amalgamated Union of Engineering Workers (AUEW).

## Craft Unions

The craft unions were based on the view that a craftsman or member of a particular trade should belong to a specific union based on their craft, irrespective of the type of industry in which they were employed. Much effort was devoted to maintaining the craftsmens' differentials over 'non-skilled' workers.

## General Unions

The general unions represent employees from many industries and skill levels but particularly the less skilled. The largest general union is the Transport and General Workers which, among other areas, has members in the National Health Service, universities, chemical, manufacturing and pharmaceutical industries.

## Industrial Unions

The industrial unions attract membership from one industry irrespective of the skills or occupations of the workers. There are few industrial unions within Britain, the National Union of Mineworkers being one, but they are more common in the United States.

## White Collar Unions

The white collar unions normally recruit non-manual workers (the manual unions being regarded as blue collar). The majority of laboratory technical staff will be eligible to join such a union. One of the interesting factors with such unions in the technical sphere is that the workers, supervisors, and possibly managers may all belong to the same union.

## Union Structures at the Workplace

Every member of a union belongs to a union branch which may be based on the workplace or on several workplaces. The branch is normally administered by a chairman, secretary, treasurer, and an executive committee. The link between the committee and the individual member is the shop steward or equivalent (laboratory representative *etc.*).

## Shop Stewards

These are unpaid representatives elected from those at the workplace, although in general there is little competition for these posts. The powers of shop stewards vary considerably. They may have the authority to conduct local negotiations or they may be limited to the role of communication link and collector of union subscriptions.

## Safety Representatives

The Health and Safety at Work Act gave unions the authority to elect safety representatives to perform specific duties at the workplace. These representatives may be the shop steward or a separate person. Their role will be considered in more detail in the Chapter on Safety.

## The Union Branch

The branch deals with such matters as membership applications, resignations, implementation of national policy locally, and makes recommendations either to District level or to the union's national conference. The branch may also elect local negotiators and delegates.

Outside of the workplace, the union will have a number of other levels of lay officers *e.g.* District, Regional (Divisional) and National Committees in addition to a full-time officer structure, at regional and national levels.

## ACAS Industrial Relations Code of Practice (1972)

This code of practice covers a number of factors relating to the activities of trades unions and the need for recognition and communication between the employer and the employees' representatives. The supervisor should ensure that the procedures agreed between the employee and unions are understood at all levels within the group.

## COLLECTIVE BARGAINING

Most of the larger organizations employing technical staff will negotiate, or at least consult with, trades unions on matters relating to their employees. They may recognize such unions for collective bargaining purposes on all matters relating to staff, if full recognition has been awarded, or in other cases only as the appropriate body to deal with matters relating to individual members. Smaller companies, particularly those in the private sector, may not recognize unions at all and insist on individual bargaining with each employee or dealing with a staff association.

Collective bargaining is the process by which an agreement on wages or conditions of service is reached between the employer and the trade union. Such agreements cover both union members and non-members and will usually deal with all aspects of employment including salaries, overtime, disciplinary and grievance procedures, grading, and health and safety. Some of these items such as salaries may be negotiated nationally, while others will be dealt with locally as may be the actual details of implementation in respect of national agreements.

Within areas of the public sector, Whitley Councils or National Joint Councils of employer and staff representatives will fulfil these functions centrally. In other areas negotiation may take place between employers' associations and groups of unions or between individual employers and one or more unions.

Whatever the method of negotiation the agreement is said to be arrived at by mutual consent and is not legally binding.

At plant, university, or research centre level the personnel section is normally responsible for negotiating with the unions but the supervisor has direct involvement in a number of areas relating to the control and work of the section.

## AGREEMENTS DEALING WITH NORMAL WORKING PROCEDURES

### Overtime

These may provide for a minimum payment, 'call-out' payment, and offer the employee the choice of payment at enhanced rates or time off in lieu. The supervisor may need to get authority from a manager where staff wish to be paid for overtime while having the authority to agree to time off in lieu. Health and safety agreements providing for a minimum of safety cover in respect of first-aiders, access to external telephones *etc.* may restrict the ability of staff to work when the plant or laboratory is officially closed.

**Shift Working**

Agreements may restrict the ability of staff to change shift or to work through rest periods.

**Changes in Job Content**

There may be a provision for an 'acting up' allowance to be paid where a member of staff assumes the responsibilities of a higher grade, or where jobs are regraded when duties change or new tasks are added.

**Changes in Procedures and Machinery**

Trades union safety representatives may have requested that they be informed of any significant changes in working practices or of new equipment or machinery introduced.

**Appointment Procedures**

Procedures for the advertising, appointment, qualification, and training of staff may be covered by one or more local or national agreements. The effective supervisor will not rely on the personnel section for ensuring that all aspects of such agreements are followed. This is particularly important in respect of technical education as lay trade union officials are often more aware of the changes in this area than administrative staff.

**Holidays and Other Leave**

In addition to approving the normal holiday entitlement of staff the supervisor will need to be aware of agreements dealing with leave for special situations *e.g.* maternity, bereavement, and public service *e.g.* magistrates, territorial army, professional meetings, and sickness.

**MATTERS RELATING TO THE CONTROL OF STAFF**

While the trades unions will have been involved with the negotiation of normal working procedures these should not usually result in direct contact between the supervisor and a representative of a trade union. In matters relating to the control of staff and safety there is likely to be direct contact with either the trade union shop steward or safety representative.

### Disciplinary Procedures

These procedures may provide for the notification of the relevant trade union official at either the informal or first formal stage. Where the supervisor belongs to a different union to the person being disciplined, it might be useful to keep their own representative informed of proceedings, in general terms, in case problems arise in the future.

### Grievance Procedures

These normally allow for the aggrieved to be accompanied by a trade union representative at all stages. In such cases, the supervisor may find that the standard fairness or ability of supervisors are given as the cause of the grievance and that a good relationship with the trade union representative can do much to ensure a sympathetic hearing.

### Safety Inspections and Investigations

The supervisor may have to justify working procedures, training methods and standards of supervision during inspection by safety representative during inspections made under the Health and Safety at Work Act. The use of a checklist system during safety audits by the supervisor or during joint inspections may reduce areas of disagreement.

The position may be more difficult where the representatives are conducting investigations of specific complaints made by subordinates as the supervisor will need to justify the procedures used while maintaining the confidence of the subordinates.

### Annual Reviews and Applications for Upgrading

While individuals may apply for upgradings on the basis of changes in job content they will probably consult with their trade union representative for expert advice in preparing their job description. This may result in areas of conflict where the supervisor is attempting to retain the job at its present grade. Where the supervisor is seeking promotion for a subordinate there may be advantages in making use of trade union advice in the wording of the job description.

Barclay, in his survey of Technical Managers, found that industrial relations involvement was low in terms of activity but high in terms of problems. Such problems were more frequent in the manufacturing sector than in the service sector.

CHAPTER 12

# Health and Safety

All supervisors will have responsibilities relating to the health and safety at work of others although these will vary considerably depending upon the organizational structure of the employing body and the precise nature of the supervisor's job. They may not reflect the individual's relative position in the hierarchical structure nor indeed in the 'safety' structure. Safety Officers and members of Safety Committees may only have an advisory function whereas the line supervisor will often be named as having executive responsibilities for specific areas or functions.

This illustrates one of the real problems of safety organization, in that the supervisor with executive responsibilities is expected to carry out increased duties in respect of safety without an increase in authority which would enable these duties to be performed. Indeed, in practice the actual authority of the supervisor has been steadily decreasing over recent years with work levels, discipline, dismissal, engagement, and many other aspects of staff control passing to managerial levels or to central specialists. This reduction in authority has in turn reduced the relative status of the technical supervisor, and in many laboratories, although the technician line supervisor will, on paper, have executive responsibilities for a given area in respect of safety, the authority will rest with a scientist or member of the academic staff who is not named on the safety policy as having any specific duties, thus representing the alternative positions of responsibility without authority and authority without responsibility.

## SAFETY OBJECTIVES

Safety should be seen as part of the normal management function of the organization with high standards of safety being considered a managerial objective pursued in the same way as other such objectives. ('Managing Safety'. The Accident Prevention Advisory Unit HMSO, 1980). The objective should involve all levels of the organization and should range from the broad strategy issued by the Board or Council to the specific rules relating to the use of a particular piece of equipment prepared by the line supervisor. These objectives should be brought together in the organization's Safety Policy—a document which every employer is required to prepare under the Health and Safety at Work *etc.* Act (1974).

## Safety Resources

The objective should be practical in nature not generalized or 'wishful thinking'. This requires that sufficient resources are made available at each level of the organization. In this context resources include not only money, equipment, and specialist expertise but time and supporting services.

There are many cases where organizations have produced detailed statements of policy, meeting the best of modern safety standards but have failed to make the resources available to match them. For instance the safety policy may state that all portable electrical equipment will be tested annually by an approved competent person without provision being made to employ additional staff to test equipment or repair 'failed' items. The technician conducting the tests, the departmental safety officer, the safety committee, and the organization's main safety adviser will be aware of the situation but the policy is not revised to reflect the actual situation. Such an unrealistic approach causes the management of safety to be questioned and would lead to problems if an untested item caused an accident.

On the whole, health and safety problems do not require a high level of expenditure. The majority of incidents arise from simple things; poor housekeeping, failure to repair the building fabric, in maintaining and updating environmental controls, and poor supervision.

The Health and Safety Executive Accident Prevention Unit (in 'Effective Policies for Health and Safety', 1980) have found that few managers have adopted effective methods of controlling the safety budget or of ensuring that the resources available are used in accordance with demonstrable priorities. The same report also indicates what may be an appropriate tactic for the supervisor seeking to overcome financial constraints in respect of safety when it states that poor publicity or criticism of an organization's performance in the areas of accident, ill health, pollution, or loss of amenity, diminishes its image not only in the eyes of its neighbours but also in those of potential customers. But the use of publicity to improve safety could result in the improvements being made at the expense of the supervisor's career.

## Motivated and Involved Workforce

The importance of a well-motivated workforce is discussed elsewhere. In respect of safety all staff need to be committed to maintaining standards and in complying with the safety policy. Subordinate staff should have a direct input into the policy as it affects their own techniques and areas of work. The views of laboratory staff on the organization's attitude to safety might be expected to have a considerable effect on how seriously they respond to safety rules and procedures.

In a survey carried out by Laboratory Equipment Digest in 1980 (LED Dec. 1980) 29% of respondents felt that there was too little emphasis on safety in laboratories with 10% feeling that the greatest obstacle to better safety was bad management.

The use of incentive or reward schemes for reductions in accidents, suggestions to improve procedures *etc.* may be of value, as a means of increasing safety awareness. Effective investigations of accidents or near misses, together with routine safety audits, are important methods in showing a commitment to safety and concern for the interests of staff.

## ACCIDENTS

Where a person is injured at work, the employer may have a duty to record the injury or to notify the appropriate authority of it. The responsibility for completing the accident report *etc.*, may be delegated under the employer's safety policy to the supervisor, or in certain circumstances, the victim. Where responsibility is delegated to the individual employee to record their own accidents it might be appropriate to make other arrangements in the case of death or major injury (!).

## Incidence of Accidents

Dewhurst (F. Dewhurst, 'Laboratory Accidents and First Aid Provision' in Laboratory Management Symposia First Aid at Work, Institute of Science Technology, London) makes the point that many laboratory accidents are not accidents in the dictionary definition of 'events without cause' but result from all too apparent causes such as poor laboratory design, bad working practices, and inadequate management structure. As such situations are all too common it is perhaps surprising that the incidence of serious accidents is not greater than it is. Harrington (J. H. Harrington *Laboratory Practice* **29**, 1980) reported that recorded accidents in British Medical Laboratories resulting in injuries occurred at the rate of 25 per 100 persons at risk. Uldall (A. Uldall *Scand. J. Clin. Lab. Invest.* **33**, 1974) reporting on accidents in Danish Chemical Laboratories gave a figure of 1.4 per 100 person years compared with 0.5 for clerks.

Kebblewhite (J. Kebblewhite *The Safety Practitioner* **2**. No. 12, 1984) cited 4.3% accidents to people at risk within ten British Universities compared with 34.2% in chemicals, 12.7% in instrument engineering and 30.5% in mechanical engineering.

The apparent low incidence of accidents in the universities may be as a result of under-reporting, the basic safety of staff, lack of pressure, or the comparatively low scale level of activity compared with industrial or pilot plant work.

Within the universities surveyed by Kebblewhite it was found that the majority of accidents involved non-technical locations and this indicates the need for laboratory supervisors to concern themselves with commercial areas outside of the immediate control if they wish to reduce the risk of injury to their staff.

Within laboratories the main hazards would appear to arise from the handling of glass/sharps (needles, scalpel blades, *etc.*) and spillages/release of chemicals. Both of these would seem to be capable of reduction by substitution of plastics for glass where possible and by improved training. The position of the supervisor in recording accidents will be discussed later.

It is not, however, adequate to record merely the accident rate, which is a negative approach measuring as it does only the 'failure' rates. If, as the HSE suggest in 'Managing Safety' an effective means of measuring safety is to be adopted, a means of assessing the way in which work hazards are eliminated or controlled is required. This needs a quantitative and qualitative view of the standards of compliance with predetermined lack of performance *e.g.* legislation, codes of practice *etc*. The Chemical and Other Substances Hazardous to Health (COSHH) Regulations 1988, requires that formal risk assessments are made of the workplace. This has added considerably to the demands made upon the supervisor. The authors are among those organizing training courses to equip laboratory staff to meet these demands.

This overview of safety must include not only all accidents and near misses but diseases and related processes. The importance of the latter is illustrated by the fact that 54.7% of the respondents gave a positive reply to the question 'Do you personally know of any cases of a laboratory worker having to give up work because of dermatitis/ allergy?', in the Dewhurst survey (Laboratory News, 1973).

## The Notification of Accidents and Dangerous Occurrences Regulations (NADOR 1980)

These require that an accident is notifiable if it results in:

(i) death or major injury, or

(ii) an employee being incapacitated for work for more than three consecutive days (excluding the day of the accident and any Sunday).

In the case of death or major injury a 'responsible person' must notify the enforcing authority by the quickest practical means followed by a written report within seven days.

For minor injuries incapacitating an employee for three consecutive days the employer must keep, for at least three years, a written report giving details of the following:

(i) the date of the accident,

(ii) the name, sex, age, and occupation of the injured person and nature of the injury,

(iii) the place where the accident took place, and

(iv) a brief description of the circumstances.

Where 10 or more persons are employed, all accidents must be recorded in an accident book. This accident book, in addition to being required by law, should also provide the supervisor with a means of monitoring the effectiveness of the company safety systems, deciding on areas where additional training is required and of assessing the effectiveness of improvements in procedures.

Special provisions for notification of accidents apply in certain areas. For example, under the Ionising Radiations Regulations 1985, notification must be made of any incident (or accident) involving the release of radioactive material in excess of given limits.

NADOR also requires that certain dangerous occurrences be

notified to the enforcing authority by the quickest practical means and
supervisors should ensure that full details as to the nature of these
occurrences and the procedures to be followed are included in the
company safety policy and displayed in such a way as to make staff
aware of them.

These special 14 occurrences include:

(2) Collapse or bursting of any vessel in which there was gas (including air) or vapour at greater than atmospheric pressure *e.g.*
autoclaves.

(3) Electrical short circuit or overload attended by fire or explosion
which resulted in stoppage of the plant involved for more than 24
hours.

(8) The uncontrolled release or escape of any substance or agent in
circumstances which . . . might be liable to cause damage to the
health of, or major injury to, any person.

(9) Any accident in which a person is affected by the inhalation,
ingestion or other absorption of any substance . . . to such an
extent as to cause acute ill health requiring medical treatment.

(10) Any case of acute ill health where there is any reason to believe
that this resulted from occupational exposure to isolated
pathogens.

Where an employee has suffered an injury as a result of a notifiable,
dangerous occurrence which is the cause of their death within one
year, the employer must notify the enforcing authority as soon as it
comes to their knowledge. Clearly this requirement will only be
effective if a means is established whereby the health of employees
injured (and retired) as a result of such occurrences is monitored. The
supervisor and colleagues of the injured party are probably in the best
position to see that the employer is informed of the position without
causing undue concern to the victim or the family.

### Safety Representative Investigations

Safety Representatives have a specific right to investigate complaints
by employees, to investigate potential hazards, and to carry out safety
inspections. Normally they are expected to give reasonable notice, but
where the investigation is following a notifiable accident or occurrence, advance notice is not necessary. Supervisors should be aware of

the procedures for ensuring that Safety Representatives and the employer's Safety Officer, are notified of such accidents or occurrences and be prepared to ensure that they are notified without undue delay. The role of Safety Representatives is discussed later in this Chapter.

No attempt should be made to 'clear-up' the scene of any occurrence, other than to take urgent steps to safeguard against further hazards, until the Safety Representatives and the inspector of the enforcing authority have had the opportunity to investigate as thoroughly as necessary.

## Management of Safety

The supervisor must accept that the elimination or control of risks is not only a function of management and specialist safety staff but an essential part of the job of every supervisor. The laboratory supervisor also provides practical expertise and continuity in respect of particular procedures and should not assume that a scientist is automatically an expert in a technique solely because of their academic knowledge.

In safety, as with other aspects of the organizational structure, effectiveness depends upon the ability of the supervisor to coordinate the activities of individuals within and outside the working group to meet the organizational objectives.

In safety matters, particularly where the organization does not have a detailed safety policy, responsibility for a particular problem may not be tested until there has been a failure *i.e.* an accident or incident. In such an event it will not be sufficient to claim that a procedure had operated without problems in the past. There should be a need to show regular and systematic concern.

If the role of management is to control work and if accidents are caused by a failure of control, it should not be an adequate defence for the manager, or supervisor, to argue that they were relying on the technical competence of those doing the work or on physical safeguards. In fact it is often the more experienced and competent worker who takes unnecessary risks *e.g.* a mature, qualified machinist who works without the guard in position on the lathe. The control and monitoring systems in respect of safety must be at least as good if not better than those used in respect of other aspects of the work.

Information as to the extent of risks involved in particular processes is important in convincing subordinates of the need to adopt safe procedures and the supervisor should take the appropriate measures to ensure that such relevant information is provided by management and brought to the attention of staff.

## Assessment of Hazards

The supervisor needs to be aware of the hazards inherent in the activities of the group so as to regularly evaluate them and maintain the relevant precautions. This is particularly the case since the introduction of COSHH. Procedures for making such assessments are considered later in this Chapter, as is the subject of loss of control. Identifying hazards may in practice be only the first of the problems. In those cases where the supervisor offers advice which is not accepted, this should be recorded as protection for the supervisor in case there is an accident in the future. If such an event does occur there will almost certainly be an attempt to apportion blame and as it is unusual for senior staff to seek responsibility for a failure to control working practices, the tendency is to blame either the injured person or the immediate supervisor.

## Registers and Records

In addition to accident records there are a number of registers and records that must be maintained under the Factories Act 1961, and a variety of Regulations, Codes of Practice, and Guidance Notes. These may be in the form of general registers, such as required under the Abrasive Wheel Regulations; records of examinations and tests *e.g.* Asbestos Regulations, Inspections of Scaffolds, examinations of plant *e.g.* Hoist or Lift, Steam-Boilers; of radioactive materials and Health Registers (in those industries and processes in which employees are required to submit to regular examination).

While some of these items, such as those applicable to Ionising Radioactives and Autoclaves are directly applicable to the laboratory situation, supervisors may feel that it would be appropriate to 'borrow' standards from other situations (*i.e.* factories under the Factory Act) and apply them to similar items or equipment within the laboratory building. Such 'borrowing' should not be confined to technical processes but should include such items as ladders (Factories Act 1961, Construction (working places) Regulations 1966). In fact it is such mundane pieces of equipment as ladders and steps which are often overlooked in laboratory safety procedures, as the tendency is to concentrate on factories relating directly to the work of the company or department.

## HEALTH AND SAFETY AT WORK ACT

The Health and Safety at Work Act (HSWA), 1974 was considered a revolutionary piece of legislation in that it not only extended statutory safety cover to almost the whole of the workforce (and indeed members of the public), but it changed the emphasis from specific dangerous procedures to the total working environment; from accidents to health, with those creating the risks being responsible for them.

The provisions of HSWA Part I (Section 1(1)), were designed to have the effect of:

(i) securing the health, and welfare of persons at work,
(ii) protecting persons other than persons at work against risks to health or safety arising out of, or in connection with, the activities of persons at work,
(iii) controlling the keeping and use of explosive or highly flammable or otherwise dangerous substances, and generally preventing the unlawful acquisition, possession and use of such substances, and
(iv) controlling the emission into the atmosphere of noxious or offensive substances from premises of a prescribed class.

Regulations and codes of practice issued under HSWA give effect to the general principles set out above and include the Health and Safety (First Aid) Regulations 1981, which within its particular area, should have had a considerable impact on laboratories.

The main section of the Act as far as it effects supervisors is section 2. This requires employers, as far as is reasonably practical, to provide:

2(2) (a) plant and systems of work that are safe and without risk to health,
(b) safety and absence of risk in connection with the use, handling, storage, and transport of articles and substances,
(c) information, instruction, training and supervision to ensure the health and safety of employees,
(d) a place of work that is safe and without risks, and safe access and egress from it,
(e) a working environment which is safe and without risks to health, and adequate arrangements for the welfare of employees.

**Limited Effect of the HSWA**

In practice, the effects of the HSWA would appear to be disappointing in many respects. Dewhurst and Goldberg (F. Dewhurst and D. J. 'Goldberg Laboratory News', 1982) compared the results of safety surveys carried out before the act was introduced (1973) and after its operation for seven years (1981). The results indicated that 65% of the individual respondents had changed working practice during that time for reasons of safety, but factors relating to the establishment and implementation of safe systems of work still left much to be desired. This was illustrated by the extent to which eating, drinking, and smoking still occurred in laboratories. These three activities, almost universally condemned as unsafe practices, serve as an indication of the organization's safety effort and the effectiveness of supervisors in enforcing basic safety procedures. They involve supervisory and safety elements only, expenditure not being required to prevent any of these practices.

Figure 12.1 shows that while improvements have been made, over 23% of laboratories would appear to lack adequate levels of safety supervision.

| Practice | 1981 | 1973 |
| --- | --- | --- |
| Eating in laboratories | 27.1 | 54.0 |
| Preparing drinks in laboratories | 23.7 | 35.9 |
| Smoking in laboratories | 23.4 | — |

**Figure 12.1** *The extent of three practices indicating poor levels of safety supervision in laboratories.*

Section 2(3) requires all employers, except for such cases as may be prescribed, to prepare and revise as appropriate a written statement of their general policy with respect to the health and safety at work of their employees, and the arrangements for carrying out that policy. It also requires that the statement be brought to the notice of all of their employees without specifying the means of doing so.

The latter point is of some importance to supervisors as some scientific establishments, with the agreement of the local HSE Inspector, consider depositing a copy of the policy in the departmental library as being an adequate method of meeting this aspect of the act. In such cases the supervisor must make a positive effort to discover the contents of the safety policy and transmit its contents to subordinate

staff. Clearly this limited method of publicity for the safety policy is not
as satisfactory as the alternatives of providing every employee with
their own copy or of ensuring that a detailed copy is provided in every
section or suite of laboratories. This is particularly important as part
of the policy should consist of a detailed document or collection of
documents including manuals of procedures, rules *etc.* to which
regular reference will need to be made.

Supervisors will not normally have responsibility for preparing the
overall safety policy but will frequently be involved in the drafting of
the rules and procedures applicable to their own laboratory or section.
Indeed it is essential that they have an input to these aspects of the
policy so as to ensure that the contents are both practicable and
sufficiently comprehensive. Guidance on the contents of Safety
Policies are contained in 'Safety Policies in the Educational Sector'.
HSC. HMSO, 1985.

## Enforcing the Law on Health and Safety

The Health and Safety at Work Act created two new bodies: the
Health and Safety Commission (HSC), responsible for developing the
law and policy; and the Health and Safety Executive (HSE), respon-
sible for enforcing the law. The HSE combined those enforcing bodies
that operated before the introduction of the HSWA, including the
Factory Inspectorate and the Employment Medical Advisory Service.
In addition certain responsibilities, such as public health, rest with the
Local Authority.

## HSE Inspections

Inspectors may visit the workplace during the course of a routine
inspection, or to investigate specific problems drawn to their attention
by individuals or trades unions. Inspections are often made without
prior notice, as this might defeat the purpose of the inspection by
allowing hazards to be removed and the site to be tidied.

The Inspector will normally only be concerned with enforcing levels
of safety laid down by statute or by recognized standards. As good
laboratory practice would normally be in excess of these standards,
inspections should present the conscientious supervisor with few
problems.

**Powers of HSE Inspectors**

Section 20 of the HSWA gives the Inspectorate powers to:

   (i) enter premises (accompanied by the police or others),
  (ii) inspect premises,
 (iii) require that areas be left undisturbed,
 (iv) collect evidence (measurements, photographs, *etc.*),
  (v) seize, render harmless or destroy items causing danger,
 (vi) take possession of articles,
(vii) require assistance.

The Inspectorate also has the power to issue improvement and prohibition notices:

*Improvement Notices:* these require an employer to make specific improvements or changes within a given time.

*Prohibition Notices:* where there is an imminent risk of personal injury, an Inspector can forbid the work to be carried out until faults are put right.

While the HSE can initiate prosecutions, it has a policy of not using criminal proceedings except in cases of repeated, deliberate or severe offences.

## LABORATORY SAFETY POLICIES

At the laboratory or section level there are four main stages in preparing, or revising, a safety policy. These are the identification of hazards, deciding on control measures, the allocation of responsibilities to ensure that the procedures work, and the training of staff.

### Identification of Safety and Health Hazards

A systematic approach to the identification of potential hazards and inadequate welfare facilities is essential. There are a number of specialized publications designed to assist in this task of which 'Safety Audits: A Guide to the Chemical Industry' (Chemical Industries Association) and 'Laboratory Safety Audits and Inspections' (IST/ Northern Publishers (Aberdeen)) are recommended.

    In attempting to identify potential hazards the supervisor should consider more than just the conventional 'laboratory safety' items of procedures, equipment, and storage. The building structure, access and egress, environmental elements (heat, noise, light, ventilation) disposal procedures, emergency procedures, welfare facilities, and arrangements for visitors/contractors should all be considered. Particular items may be hazardous to specific groups of staff, for instance there are a number of risks to pregnant women such as radiation, antibiotics, certain chemicals, and anaesthetics which can cause spontaneous abortions.

The identification of a possible hazard may in practice be only the first of the supervisor's problems. It must be accepted that the interests of various groups may vary with differing priorities between groups which will affect the willingness of managers to accept that particular items are hazardous.

## The Elimination or Control of Hazards

Once a hazard has been accepted as such the first objective should be to remove it by changing procedures or equipment. Unfortunately this is often not possible as the hazard results from the very nature of the work, such as infections from bacteria in microbiology laboratories, or it is held to be not reasonably practicable to do so, for example by being too expensive in view of a limited risk. In such cases the policy should be to reduce or control the risk by: changing working methods, improving equipment, training staff, permit to work systems, increased supervision and/or the use of protective equipment. Permit to Work Systems may have a role to play in controlling risk and are discussed later in this chapter.

## Training

Although the Act places the responsibility for the training of staff on the employing organization, relatively few changes have been made to the existing pattern of technician training to meet this obligation. Figure 12.2 indicates the main areas of training necessary to meet the provisions of the Act.

*Training in Laboratory Techniques.* Traditionally, training in the use of laboratory techniques and procedures provided as part of the formal educational programme has included elements relating to their safe use. These items have been supplemented with additional informal training, provided at work, in specific skills and their application. At the bench level this would appear, on the basis of accident figures, to

(1)  Induction training for new recruits including introduction to safety rules and policy.

(2)  Updating training for all existing staff on techniques, procedures and standards.

(3)  Pre-promotion training so as to ensure that all promoted or transferred staff receive training in their new tasks before assuming responsibility for them.

(4)  Training in instructional, supervisory and managerial skills.

(5)  Ensure that safety representatives are adequately trained.

(6)  Record training (i)   provisions in the safety policy,
     and (ii)  given and received by individual members of staff.

**Figure 12.2**  *Aspects of training required to meet the requirement of HSWA.*

have been reasonably effective, but it is probably not adequate to meet the requirements of the act. The Health and Safety Executive recommend that there should be a proper assessment of training needs at all levels within the organization. This would require supervisors and managers, in consultation with specialist staff, to prepare a specific training programme to meet the needs of all new entrants to each laboratory or section. The addition of regular up-dating and refresher courses should be arranged both routinely and as the law and regulations change. Such programmes should not be limited to those specialized areas currently recognized as warranting particular attention *e.g.* radiation, but should cover all aspects of the work, *e.g.* noise, dust, equipment, machinery, toxic hazards, storage, and fire.

*Supervisor Training.* The absence of specific training for laboratory supervisors is a subject of some concern both in respect of their duties under the organization's safety policy and to enable them to deal effectively with trade union safety representatives. Often as a result of lack of training and information supervisors feel that safety legislation erodes their status whereas used properly by management it can increase their authority. Supervisors' training should cover the five main areas as indicated in Figure 12.3.

In respect of the ability of supervisors to deal effectively with Safety Representatives there are five main areas in which expertise is required (Figure 12.4). One of the main problems relating to these items is that the employer may not provide such training. Where this is the case the supervisor should undertake the task of equipping himself to

**Organization Safety policy and the system by which it is implemented**
(1)   Policy at central and departmental levels.
(2)   Supervisors responsibilities and authority, safety specialists.
(3)   Safety structure, role of trades union representatives.

**Hazards**
(4)   Recognition of hazards relating to specific work areas audits and inspections.
(5)   Elimination or control of hazards.
(6)   Communication systems for reporting hazards.

**Staff**
(7)   Recognition of training needs—work related and general.
(8)   Basic training skills and instruction technique.
(9)   Protective clothing and equipment.
(10)   Enforcement of rules and communicate skills.

**General**
(11)   Organizations fire and first aid procedures..
(12)   HSWA, relevant codes of practice *etc.*
(13)   Special systems as appropriate *e.g.* permit to work.

**Figure 12.3** *Basic syllabus for supervisor training in safety.*

(1)   Ability to conduct safety audits and inspections of processes and equipment.
(2)   Knowledge of rules, regulations and procedures governing use and storage of chemicals and materials used.
(3)   Knowledge of safe systems of work appropriate to the areas under their control.
(4)   Knowledge of the cost of improving any items that do not meet the necessary standards.
(5)   Understanding of the code of practice in respect to safety representative and any company agreements governing their operation.

**Figure 12.4** *The basic requirements of supervisors in dealing with trade union safety representatives.*

deal with the matter on his own initiative. LSTS organizes distance learning courses to meet this need.

### Responsibilities for Ensuring that the Safety Arrangements Work Effectively

This aspect of the safety policy is the one which may give rise to difficulties between the supervisor and the manager acting on behalf of the employer. Arscott, writing in 'An Employer's Guide to Health and Safety Management' (Kogan Page for the Engineering Employers' Federation, 1980), states that supervisors have the 'onerous responsibility to balance the need for high output and quality, and low costs, with the need to ensure that the system at work is safe'. He further suggests that they have specific duties to instruct employees in safe working and to enforce safety rules, to provide constant supervision of staff and facilities, to investigate and take action following incidents, and make recommendations. Under such a system, management works through the supervisors to implement the health and safety policy. This is as would be expected under a normal organizational structure, but while specific duties may be allocated to individual supervisors overall responsibility should remain with the employer. However in many organizations there would appear to be a deliberate policy on the part of senior staff to delegate many of their responsibilities down to line supervisors. This desire to protect their own interests on the part of management is understandable in the light of possible prosecution under the act, but it should not pass unchallenged. Figure 12.5 illustrates this point with a major part of the Head of Department's responsibilities being delegated to supervisors. It is unlikely that supervisors would in practice have the means or authority to carry out these responsibilities.

This is an area of considerable importance to the supervisor as there may be a risk of prosecution under the HSWA, and more importantly it effects the degree and nature of supervision provided. In many laboratories, the safety policy may provide the only formal information available to the supervisor on the organizational structure of the establishment, and some aspects of such policies will be considered before covering the wider implications of the act.

### Supervisor's Approach to the Safety Policy

The safety policy is clearly a most important series of documents which should be examined carefully by every supervisor. The trade union, the Association of Scientific Technical and Managerial Staffs (now the Manufacturing Science and Finance Union) in 'Health and Safety and the Supervisor' (1981) recommend that supervisors bear in mind the following principles:

   (i) All health and safety duties must be clearly defined.
  (ii) The duties should be part of the normal routine of the job and not

increase the workload to such an extent that they cannot be performed.

(iii) The company should provide adequate training and support facilities.

(iv) Permit to work systems should be examined closely to ensure that the limitations and restrictions are enforceable.

(v) The procedures contained in the policy should be practicable, produce safe systems of work, and be enforceable.

**Prosecution of Supervisors under the HSWA**

There is a degree of concern among supervisors regarding the risk of prosecution under HSWA and this concern is encouraged by some employers in an attempt to give strength to their safety organization. There are two main ways in which this is done. The subtle approach consists of wording the safety policy as discussed in the preceding pages with reference to 'responsibilities' and 'requirements' under the act. In other cases a more robust style may be adopted with the employer issuing a circular to the effect that 'Supervisors have always been responsible for the safety of their technicians, and the new legislation makes failure to take reasonable care to ensure their safety a criminal offence.'

In fact, the vast majority of prosecutions are against the employing organization or in a small number of cases against senior executives (under section 37(6)). The offence of failing to take reasonable care relates to all employees and not specifically to supervisors. In such cases it was felt that the Inspectors would only consider prosecuting individual supervisors if they were convinced that the employer and all levels of management had done everything they could to fulfil their duties.

However, there has been at least one case where supervisors have been charged and pleaded guilty under Section 7 of the Act. The supervisors knew of a fault in an item of equipment which resulted in an accident but had failed to report it. In this case it was argued that the offence was of an omission to report directly to the maintenance section that part of a machine's essential design for safety (an interlock on a door) was at fault, there being no reason to suspect that the firm would not have taken reasonable action to deal with the problem. In another case a science teacher, as the area supervisor, was fined after an accident in a class demonstration. The employer was not prosecuted.

**General Statement of Safety Policy**

Heads of Departments and Institutions are responsible for the health, safety, and welfare of all the people who are lawfully in the buildings within their charge.

**Departmental Statement of Safety Organization**

As Head of Department I am responsible for supervising the Safety Policy in respect to this department. I have delegated some of these responsibilities as described in this document.

Every supervisor is responsible for ensuring as far as is reasonably practicable the safety of staff, students and other persons in their area. In particular the following responsibilities are delegated to each supervisor.

(a)  To ensure that **all** procedures used are safe and in compliance with **any** relevant code of practice and that instruction in safe practice is given.

(b)  To inform the Safety Officer of any special or newly identified hazards or of new hazards about to be introduced.

(c)  To post prohibitive and warning notices and signs. To ensure that such signs are obeyed at all times.

(d)  To ensure that fire escape routes are kept clear.

(e)  To ensure the safe disposal of noxious waste in compliance with local regulations.

(f)  To ensure that all highly flammable liquids are stored in accordance with Statutory Regulations.

(g)  To ensure that all staff, students and visitors are adequately trained in procedures. To ensure that all visitors are fully informed of emergency procedures.

(h)  To ensure that all contractors work in accordance with the safety policy and that no work is carried out which is likely to endanger company employees.

**Figure 12.5** *Extracts from a safety policy indicating how management may attempt to shift responsibility for safety onto supervisors.*

## An Example of an Employer's Safety Policy—Figure 12.5

The introductory paragraph to the supervisor's duties in Figure 12.5 may at first glance seem reasonable enough. All supervisors should try to ensure that the areas under their control are as safe as possible. However, a number of points are raised by such general statements.

The first relates to the use of the word 'responsible'. The point has already been made that the act places responsibilities on the employer and it may be argued that duties rather than responsibilities should be delegated to the supervisor. Factors relating to authority raise problems of a more practical nature; does the supervisor have sufficient authority to ensure that all 'persons in their area' comply with their wishes? How would management react if a line supervisor attempted to make a visiting senior scientist comply with the section's safety precautions?

The supervisor attempting to comply with such duties would also need to be aware of the disciplinary procedure for dealing with staff who do not comply, and whether any system exists for disciplining the supervisor who is found to be unable to enforce safety procedures. In conclusion such statements are not sufficiently specific as to be of any value as they define neither the duties nor authority.

(*a*) *Safety of Procedures.* Section (a) states that 'To ensure that all procedures used are safe and in compliance with any relevant code of practice'. Statements of this kind actually reverse what should be considered as normal managerial practice. It is the duty of the employer, not that of the supervisor, to ensure that safe systems of work are used, to provide hazard information *etc.* No procedure, product, chemical or machine should be introduced or used without the supervisor requesting written instructions on its safe use in the particular circumstances applicable to that section. In practice, management may lack the expertise to provide such information and in such cases the laboratory supervisor should prepare and submit draft procedures for the approval of management. Any Codes of Practice to be followed should be named in the safety policy and copies supplied for the use of the supervisor, along with details of local variations and additions, such as precise details relating to disposal of chemicals.

(*b*) *Hazard Information.* 'To inform the safety officer of special or newly introduced hazards'. In certain circumstances, a variation on this sentence could be useful. If as a result of an experiment or procedure a new hazard is discovered then clearly the supervisor should notify the Safety Officer at once. However, the paragraph in Figure 12.4 goes considerably further than such commonsense provisions and attempts to make the supervisor responsible for notifying management of all new hazards introduced. In general it should be management who inform the supervisor that a new chemical or procedure is hazardous before the procedure is introduced.

(*c*) *Notices and Signs.* Most such signs will be related to the use of safe systems of work and as such should be provided by management. The signs may concern statutory duties, management rules, or may provide information and should be reviewed before the supervisor accepts such duties. When signs are displayed but regularly ignored, management should be informed in writing. Enforcement may again produce difficulties which relate to the authority of the supervisor.

(*d*) *Escape Routes.* It is reasonable to delegate responsibility to the supervisor to ensure that fire escape routes are kept clear as long as the duty only applies to the area under the supervisor's immediate control. Responsibility for corridors may be delegated to the maintenance section under the safety policy and any areas of overlap should be clearly defined. Line Supervisors should not accept responsibility for ensuring that staff are trained in, or aware of, the fire evacuation procedure and assembly points unless such duties form part of a coordinated training programme covering all sections of the establishment, with the procedures being capable of use in the case of a major fire.

(*e*) *Disposal of Noxious and Other Waste.* Traditionally the disposal of waste has formed part of the educational courses for science technicians. However, in practice this duty hinges on the provision of information relating both to the specific hazards associated with such types of waste, with each chemical used, and the local rules applicable. A supervisor cannot be expected to dispose of waste safely and in accordance with rules and regulations unless written details of these are provided by management.

(*f*) *Storage in Accordance with Statutory Regulations, Codes of Practice etc.* Once again the ability of the supervisor to comply with such duties is dependent upon management providing copies of the relevant statutes and authority to purchase the necessary storage cabinets *etc.* This also raises the problem of enforcement if such storage facilities are not provided near to the area where the chemicals *etc.* are to be used.

(*g*) *Training of Staff, Students and Visitors.* Where staff training forms part of the normal duties of the supervisor it would seem appropriate that training in safety procedures be included as part of these duties. However, adequate information as to hazards *etc.* must be provided by management and supervisors should have any training syllabuses they prepare officially approved. Technical supervisors will rarely have sufficient authority over students and visitors to comply with any requirements to train them. In the case of the latter, the responsibility for ensuring that they are informed of emergency procedures can only rest with their host, and it is unrealistic to expect the supervisor to pounce on every stranger passing through their area and inform them of the emergency procedures.

(*h*) *Contractors on Employer's Premises.* It is clearly unrealistic to expect individual supervisors to accept responsibility for the conduct of outside contractors. The technical supervisor will not have the time, knowledge, expertise or authority to perform such duties. The person responsible for placing the contract should provide contractors with a copy of the relevant sections of the safety policy and arrange for a 'Clerk of Works' to monitor compliance. More detailed monitoring is outside of the scope of this book and a useful summary is contained in Safety Digest No. 5 published by the Universities Safety Association (November 1984).

### Specialists and Others with Advisory Functions

The safety policy may make named individuals responsible for specific aspects of safety over the whole building, works or campus. Such responsibilities often relate to aspects of electrical safety, radioactive substances, pathogenic micor-organisms or solvents. In such cases the relationship between such specialists and the line supervisor will need to be clearly defined, particularly in respect of authority and responsibility if an item relating to a specialist function is misused in the supervisor's area.

### Information from Suppliers and Manufacturers

The HSWA places an obligation upon suppliers and manufacturers to ensure that:

   (i)  articles are of safe design and construction,
  (ii)  articles are adequately tested, and
 (iii)  adequate information is available to the user.

These duties do not detract from those of the employer to ensure the health and safety of employees. The user management *i.e.* the laboratory, must still comply with the requirements of subsection 2(2)(C) in ensuring that the ultimate user at the bench is provided with such information, instruction, and training as is necessary. It is in these areas that the supervisor will have responsibilities but under some employer's safety policies the supervisor may have the additional duty of providing the safety information for the bench worker.

As is explained in the HSE Guidance Note (GS8) and is indicated in (iii) above manufacturers, importers and suppliers of substances have a duty to ensure that adequate information about the products they supply will be available to the user. It means that the user should have access to the information, not that every substance supplied should be accompanied by information. A certain amount of information may in fact be always provided with the substance by means of a label carrying primary warnings and instructions. In most cases there will be a need for more information than could fit on a label, or in a catalogue, and the use of Data Sheets provide a means of overcoming this difficulty. Although it is obviously desirable that such sheets are provided automatically they are frequently supplied only upon request.

Supervisors responsible for ordering substances and equipment should consider routinely including a request for data sheets or safety information on every order. Where an item is to be used for a novel purpose the end use should be indicated as it would be unreasonable to expect the supplier to predict all the uses to which a substance may be put. In such cases consultation between the employer and the supplier would be necessary to ensure that the information supplied is adequate under each set of circumstances.

Unfortunately, in some instances it will be found that the information from suppliers is inadequate and the supervisor will need to inform management that this is the case and seek instructions as to the appropriate course of action. Under no circumstances should the supervisor order subordinates to use substances or equipment without adequate information or training, or work systems without direct instructions from management.

**Permit to Work**

The use of a 'permit to work' system is specifically required by some Statutes and Regulations. These require that a specific permit to work be issued to each worker authorizing them to undertake a particular process or certifying that it is safe for them to do so.

The Chemical Works Regulations 1922 provide an example of such a system in that they require a responsible person appointed by the occupier to record in a book specially kept for the purpose that it is safe for an employee to enter any place where there is reason to apprehend the presence of dangerous gas or fume. Within laboratories, such systems may be introduced to meet the employer's duties under S2 of the HSWA where there are particularly hazardous processes. Examples of such situations are work on the electrical distribution system, research work with high voltage apparatus, modification to electrical equipment, work involving the use of carcinogens, work with radioisotopes, and work with dangerous pathogens.

The Factories Act (Section 31 (4)) required that any tank or vessel which has contained fumes of a flammable or explosive character may not be subjected to welding or any process involving the application of heat until all practicable measures have been taken to remove traces of fumes or to render it non-explosive. This requirement could usefully be 'borrowed' by any supervisor responsible for laboratory mechanical engineering workshops and would appear to be suitable for a permit to work system.

Where a permit to work system is in use there should be established procedures to ensure that all those at risk, including contractors and other occasional or unexpected users of equipment *etc.* are made aware of the system.

The supervisor will need to ensure that the system is practicable, and easy to follow.

## SAFETY COMMITTEES

Supervisors are likely to be invited to serve as members of Safety Committees at both company (or works) or departmental (or section) level. It is most important that the supervisor clearly understands the terms of reference of the committee before accepting such a position.

Safety committees may be advisory in function or have executive powers, they may be confined to management representatives only or be joint managerial/employee committees.

The Safety Representatives and Safety Committees Regulations 1976 (SRSC) prescribe the cases in which an employer shall establish a safety committee as follows:

9 (1) For the purpose of section 2 (7) of the 1974 Act (which requires an employer in prescribed cases to establish a safety committee if requested to do so by safety representatives), the prescribed cases shall be any cases in which at least two safety representatives request the employer in writing to establish a safety committee.

(2) Where an employer is requested to establish a safety committee in a case prescribed in paragraph (1) above he shall establish it in accordance with the following provisions:

(a) he shall consult with the safety representatives who made the request and with the representatives of recognized trade unions whose members work in any workplace in respect of which he proposed that the committee should function;

(b) the employer shall post a notice stating the composition of the committee and the workplace or workplaces to be covered by it in any place where it may be easily read by the employees.

(c) the committee shall be established not later than three months after the request for it.

The regulations also include as one of the functions of Trades Union Safety Representatives (Regulation 4.1(h)), 'to attend meetings of safety committees where he attends in his capacity as safety representative'. From this it will be seen that not all safety committees meet the requirement of the regulations, or the HSWA, and supervisors are advised to look carefully at the terms of reference of those committees which exclude employee representation.

The guidance notes issued with the regulations make the point that 'there may be a place for safety committees at group or company level for larger organizations' but that they are 'most likely to prove effective where their work is related to a single establishment rather than a collection of geographically distinct places'.

In practice, supervisors, unless they have a particular expertise, are more likely to be able to contribute effectively to a departmental safety committee than to one with a wider remit. In the latter case one would expect the presence of senior management rather than of supervisors.

## Objectives and Functions of Safety Committees

The objectives given below are taken from the guidance notes and apply to joint safety committees established under the regulations. Departmental or 'management only' committees may have more restricted terms of reference.

7 Within the agreed basic objectives certain specific functions are likely to become defined. These might include:

(a) The study of accident and notifiable diseases statistics and trends, so that reports can be made to management on unsafe and unhealthy conditions and practices, together with recommendations for corrective action.

(b) Examination of safety audit reports on a similar basis.

(c) Consideration of reports and factual information provided by inspectors of the enforcing authority appointed under the Health and Safety at Work Act.

(d) Consideration of reports which safety representatives may wish to submit.

(e) Assistance in the development of works safety rules and safe systems of work.

(f) A watch on the effectiveness of the safety content of employee training.

(g) A watch on the adequacy of safety and health communication and publicity in the workplace.

(h) The provision of a link with the appropriate inspectorates of the enforcing authority.

It must be remembered that it is normally management who are responsible for taking executive action and ensuring the safety of employees. The role of the safety committee should be to supplement these arrangements; it cannot be a substitute for them. It is particularly important that the supervisor does not allow a reference to the safety committee to delay matters where urgent action is required, or to accept the recommendation of a non-executive committee or officer not to take action on a potential hazard without referring the matter to management in writing.

## SAFETY REPRESENTATIVES

In addition to extending the cover of safety legislation to almost the whole of the workforce, the other major initiative of the HSWA provided for employee involvement through the appointment of trade union safety representatives. While providing employees representatives with a practical involvement in safety matters, it does present areas of potential conflict for supervisors. For safety has now become one of the main areas of industrial relations and perhaps the major area in which supervisors will be directly involved with trade union representatives within laboratories.

The position is made more difficult for the supervisor by the simple fact that the majority of trade union safety representatives will have attended a ten week TUC day-release training course while the majority of supervisors will have received little training.

The Chief Inspector of Factories, Mr J. Hammer, in his 1977 report records an occasion when a 'senior' supervisor was reduced to near despair by his inability to talk on equal terms with a safety representative who had established a comprehensive library of reference material with which to support his arguments. But the problem is not simply one of facts relating to hazards, the supervisor needs to be fully aware of the legal rights of safety representatives and of any company agreement relating to their operation.

### Appointment of Safety Representatives

The SRSC regulations allow for the appointment of safety representatives by recognized trade unions *i.e.* certified independent unions recognized by the employer for the purpose of negotiation on behalf of employees.

Safety representatives will be expected as far as is reasonably practicable to have either been employed by the employer throughout the preceding two years or have had at least two years experience in similar employment. This provision is designed to ensure that representatives have adequate knowledge and experience to perform their functions.

### Functions of Safety Representatives

The Safety Representatives have functions under both the HSWA (section 2(4)) to represent the employees in consultation with the employer in respect of health and safety at work and under the Regulations. It is important to note that these are functions and not responsibilities or duties, and as such are different from those that may be delegated to supervisors.

Functions of safety representatives:

4 (1) In addition to his function under section 2(4) of the 1974 Act to represent the employees in consultation with the employer under section 2(6) of the 1974 Act (which requires every employer to consult

safety representatives with a view to the making and maintenance of arrangements which will enable him and his employees to cooperate effectively in promoting and developing measures to ensure the health and safety at work of the employees and in checking the effectiveness of such measures), each safety representative shall have the following functions:

(a) to investigate potential hazards and dangerous occurrences at the workplace (whether or not they are drawn to his attention by the employees he represents) and to examine the causes of accidents at the workplace;

(b) to investigate complaints by any employee he represents relating to that employee's health, safety or welfare at work;

(c) to make representations to the employer on matters arising out of sub-paragraphs (a) and (b) above;

(d) to make representations to the employer on general matters affecting the health, safety or welfare at work of the employees at the workplace;

(e) to carry out inspections in accordance with Regulation 5, 6 and 7 below;

(f) to represent the employees he was appointed to represent in consultations at the workplace with inspectors of the Health and Safety Executive and of any other enforcing authority.

(g) to receive information from inspectors in accordance with section 28(8) of the 1974 Act; and

(h) to attend meetings of safety committees where he attends in his capacity as a safety representative in connection with any of the above functions; but without prejudice to sections 7 and 8 of the 1974 Act, no function given to a safety representative by this paragraph shall be construed as imposing any duty on him.

(2) An employer shall permit a safety representative to take such time off with pay during the employee's working hours as shall be necessary for the purposes of:

(a) Performing his functions under section 2(4) of the 1974 Act and paragraph (1) (a) to (h) above;

(b) undergoing such training in aspects of those functions as may be reasonable in all the circumstances having regard to any relevant provisions of a code of practice relating to time off for training approved for the time being by the Health and Safety Commission under section 16 of the 1974 Act.

In this paragraph 'with pay' means with pay in accordance with the Schedule to these Regulations.

## Inspections of the Workplace

Under Regulations 5 and 6, safety representatives may conduct inspections of the workplace at three-monthly intervals, more frequently if there has been a change in conditions of work, if new machinery has been introduced, or if new information has been provided by HSE relating to hazards. Reasonable notice must be given in writing of their intention to do so.

Supervisors may be placed in an invidious position, if although notice has been given to the 'employer' of the intention to inspect they, the supervisors, have not been informed of the inspection. In such circumstances it would clearly not be conducive to good industrial relations for the supervisor to attempt to delay or prevent the inspection or to appear less than helpful to the representatives.

The HSC are of the view that joint inspections by safety representatives and management would be advantageous but this does not preclude safety representatives acting independently. In such cases the employer or manager has the right to be present during the inspection if they so desire.

## Investigations

Safety representatives have the right to investigate potential hazards and complaints by employees without giving the employer notice of their intentions. It is these rights (under 4(1)(a) and 4(1)(b) of the SRSC regulations) which may provide the greatest areas of potential conflict between safety representatives and supervisors. Faced with safety representatives wishing to carry out an investigation without notice the supervisor may be caught off guard, feel annoyed that a subordinate has complained, or be defensive at being indirectly criticized by the representative investigating one of the supervisor's procedures. As a result of one or more of these feelings the supervisor may over react, particularly if the representative is armed with factual information relating to the hazard which has not been made available through the managerial structure.

## The Right to Information

Safety representatives, provided they have given the employer reasonable notice, are entitled to inspect and look at copies of any documents relevant to the workplace of the employees represented. This information includes technical reports, details of work systems, and materials used. It does not include information affecting national security, information which would result in the contravention of a prohibition notice, cause substantial injury to the employer's undertaking or related specifically to an individual (unless the individual has consented to it being disclosed).

Supervisors should seek guidance as to what information they are authorized to give to representatives personally and what should be supplied by those higher in the management structure.

## Time Off for Safety Representatives

Representatives are entitled to time off work in accordance with regulation 4(2). The procedure governing the amount of time provided and the means of notifying the supervisor will normally be agreed as part of the normal bargaining process between the trades union and the employer. Where a supervisor has a subordinate appointed as a safety representative there will be a need for the supervisor to be familiar with the agreement and the procedures governing its implementation.

## Supervisors as Safety Representatives

In a large number of work places, supervisors will have the opportunity to become safety representatives—indeed their technical expertise, knowledge of communication skills and familiarity with the procedures for conducting safety audits would seem to make them an obvious choice. In addition, the training and information they would receive as a safety representative could be of considerable value to them in carrying out their duties as a supervisor.

While it may seem convenient for a supervisor/safety representative to inspect the areas for which they are responsible as a supervisor, it is easy for them to miss potential hazards. The old adage 'familiarity breeds contempt' applies as much to safety procedures as to other items. This may be compounded by the fact that while wearing the 'safety representative's hat' there are no legal responsibilities but once that is removed and the supervisory role assumed, responsibilities follow.

The second difficulty relates to the attitude of the employer to the activities of trade unions in general and safety representatives in particular. Where a certain type of manager is involved it may not be in the supervisor's long term interest to be too critical of the establishment. Obviously it is deplorable if supervisor's careers are inhibited by their interest in legitimate trades union activities but such situations do arise and few supervisors will be prepared to risk long term career prospects in this way.

## SAFETY STANDARDS

One of the difficulties facing the conscientious laboratory supervisor in attempting to prepare safety rules for adoption in the area under his control is the number of sources of possible safety standards (Figure 12.6). These range from legislation, which provides standards having

| Standard | Source | Status |
|---|---|---|
| Statute | Parliament | Law |
| Regulations | Minister of State | Law |
| Local Authority Regulations | L.A. | |
| Codes of Practice | HSC | Not law but acceptable as evidence |
| Guidance Note | MSE | Not law but authoritative |
| Health and Safety at Work booklets | HSE | Guidance to good practice |
| Recommendations Codes of Practice Standards | Various eg BSI Professional bodies I.L.O. | Not law but may be 'borrowed' to provide standard |
| Employers rules and recommendations | Individual employers | Enforcable through disciplinary procedures |

**Figure 12.6** *Sources of safety standards available to the supervisor.*

the full force of law, to recommendations produced by official, semi-official bodies, pressure groups, and vested interests which range in authority from providing guidance; to good practice; to discussion papers seeking to promote change. This range of sources and status of material makes it essential for the supervisor to be fully aware of the precise authority of any document being used to support a case for introducing new safety rules at work.

**Legislation Standards**

In general, statutes affecting the health, safety, and welfare of people at work may be considered under three main headings.

(1) *Safety Law*
  (i) Acts and Regulations directly relating to safety *e.g.* Health and Safety at Work *etc.* Act 1974, Notification of Accidents and Dangerous Occurrences Regulations 1980, or Chemical and Other Substances Hazardous to Health 1988.
  (ii) Acts and Regulations relating to the safe use of specific items or materials. These may relate directly to laboratory items *e.g.* the Ionising Radiations Regulations 1985, or to general items used or found in laboratories *e.g.* Asbestos (Licensing) Regulations 1983 or the Petroleum Consolidation Act 1928.

(2) *Employment Law.* The law relating to the employment of staff part of which may contain safety health and welfare provisions *e.g.* Offices, Shops and Railway Premises Act 1983. Chapter 13 deals with this area in some detail.

(3) *Other Legislation.* Although not directly related to safety or employment, other legislation contains provisions affecting matters concerned with such items as the Public Health (Drainage of Trade Premises) Act 1937 which controls the discharge of effluents into the public sewers.

It is important that all the relevant statutes are considered when deciding on an appropriate procedure to deal with a particular problem in the laboratory. A number of current statutes and Codes of Practice are appended at the end of the book.

CHAPTER 13

# The Law and the Supervisor

## INTRODUCTION

In the efficiently organized laboratory, particularly within larger organizations, the supervisor will not be required to become involved in the legal aspects of employment. Matters of this kind will have been considered by experts in the Personnel or Administrative Sections and working procedures devised to meet both the letter and spirit of the law. For instance specimen advertisements will be drawn up which comply with the Sex Discrimination Act 1975 and the Race Relations Act 1976.

This chapter does not attempt to cover in any detail the vast expanse of law as it affects those at work, but gives a brief introduction to the topic as a means of indicating how the law impinges on all areas of the employment scene. The latter is of prime importance; the law does not form a separate subject area for supervisors but forms an integral part of the background against which all supervisory activities may be seen. This is especially true in the case of what is known as 'Employment Law'.

### Common Law

The common law is a system of law based upon the acceptance of precedent, 'stare decisis', which developed over the years from the Norman Invasion. The solution offered by a judge to a particular problem in the case before him was accepted and followed in subsequent cases. Common law therefore was 'judge made', law originating from medieval customs and the system of feudal tenures.

In practice, Statute Law is of greater importance to the supervisor within the employment field although the common law duties of employees are very relevant to the contract of employment.

## Statute Law

Statute law is normally defined as the law resulting from an 'Act of Parliament' although legislation dates back to before the time of parliament in the form of Royal Charters, Assizes, Constitutions and Provisions.

Parliament has become an important source of law since the nineteenth century and of employment law in the last fifty years. Statutes are the ultimate source of law, prevailing over and in some cases abolishing common law rules.

## Delegated Legislation

Many statutes contain only general provisions, delegating authority to a specified person, or body, to issue detailed regulations which become law as delegated or subordinate legislation. The Social Security legislation provides an example of an Act under which regulations are approved by the appropriate Minister, while the Health and Safety at Work Act provides an example of legislation which allows for the subsequent adoption of 'Codes of Practice'.

## Criminal Law

The criminal law is involved when a legal proceeding is instituted on behalf of the state. In such cases the objective is to punish the guilty party or to deter him from repeating the offence against society. The chapter on Safety showed that such actions are not confined to individuals. In respect of employment, the ability for the state to prosecute companies, but not 'Crown' organizations *e.g.* Civil Service establishments, NHS, is of considerable importance. Crown immunity is provided for these government organizations as it is felt that the state should not prosecute itself. While it is said that in most areas of employment *e.g.* safety legislation, Crown laboratories will meet the standards required of normal employers it might be thought unfortunate that provision could not be made to expose the managers of such laboratories to the full force of law.

Criminal cases are normally hard by the Magistrates Court or the Crown Court before a jury, with appeals being made to the Court of Appeal (Criminal Division).

**Civil Law**

The civil law is that law in which the parties involved do not include the state. In general this entails two parties, the 'plaintiff', who is bringing proceedings, and the 'defendant'. In cases resulting from accident, dismissal or other work related actions the defendant is usually the employing organization while the plaintiff or complainant is the individual concerned.

The objective of civil proceedings is to:

 (i) restore to the plaintiff what was his, for example his job or his property,
 (ii) obtain 'damages' as recompense for suffering or injury,
(iii) obtain an injunction to stop a person (or persons) doing a hurtful act such as striking or picketing.

Cases of 'Tort' are part of the civil law. They are brought by people who have suffered as a result of someone's wrong actions and usually take the form of a claim for damages. An action in tort can only be brought where the harm suffered is caused by a wrong *i.e.* is a violation of a right which the law invests in the plaintiff. The mere fact that one has suffered injury or damage does not entitle him to bring an action.

In contrast to the criminal law, the Crown is in general liable in the same way as the public, by the Crown Proceedings Act 1947, and this enables the majority of employees to take action against their employer. Members of the Armed Forces are the major group excluded from doing this.

**Vicarious Liability**

Although the individual who commits a tort is always liable, sometimes another person may also be held responsible even though they did not themselves commit the act. In such cases they are said to be 'joint tortensory'. This is the basis of vicarious liability under which a master, or employer, is liable for the torts committed by his servant, or employee, in the course of his employment.

It is now necessary for employers to insure against this risk, under the Employers' Liability (Compulsory Insurance) Act 1969, in respect to injuries caused by his employees to fellow employees.

Cases of civil law are normally heard in the County Courts with Appeals being made to the Court of Appeal (Civil Division) and if allowed to the House of Lords.

## Employer's Negligence

Where an employee brings a case against his employer on the basis of negligence at common law it will be necessary to prove that the injury received was a result of the employer's (master's) breach of his duty of care.

The case of Wilsons and Clyde Coal Co. v English, (1938) established four areas of care; to provide:

(i) proper and safe plant and appliances for the work,
(ii) a safe system of work with adequate instruction and supervision,
(iii) safe premises,
(iv) competent staff.

It will be seen that those are not dissimilar from those given in Section 2 of the Health and Safety at Work Act but the common law duty remains.

The duty of care was defined by Lord Atkin in 1932 when he said

'You must take reasonable care to avoid acts or omissions which you can reasonably foresee would be likely to injure your neighbour. Who then is my neighbour? The answer seems to be persons who are so closely and directly affected by my act that I ought reasonably to have them in contemplation as being affected when I am directing my mind to the acts or omissions which are called in question'.

## Contributory Negligence

Often when an accident has occurred it will be found that both parties have been negligent under the doctrine of contributory negligence. In such cases the damages may be reduced according to the degree of the fault of the complainant, which is often expressed in percentage terms.

## Employment Law

Employment law is that law affecting the employment relationship and includes common law, statutes, tribunal and court decisions. It is of the greatest importance to managers, supervisors and trades unions. A number of publications have been produced on the subject. One of the most reasonably priced is 'Tolley's Employment Handbook' by E. Slade (Tolley Publishing Co. Ltd.) while 'Croner's Reference Book for Employers' (Croner Publications Ltd.), supplied in a loose leaf format, provides a regular amendment service.

## Employment Contract

Before employment law can be applied it must be established that the individual is actually employed. This is easy where a clear 'contract of employment' (or contract of service) exists but in other cases it may be difficult to decide whether a person is working as an employee or under a 'contract of services' as an independent contractor. Normally such complications will not arise in the fields covered by this book.

## Employment Legislation

The Employment Protection (Consolidation) Act 1978, as amended by the Employment Act 1980 and the Employment Act 1982, consolidated the main statutory employment rights.

ACAS, the Advisory Conciliation and Arbitration Service, was established under the Employment Protection Act 1975 to bring about an improvement in industrial relations and to improve the system of collective bargaining. To this end it has introduced Codes of Practice on Disciplinary Practice and Procedures, Disclosure of Information and Time Off for Trade Union Duties.

In the following pages we will briefly indicate those statutes relating to the various aspects of employment in approximately chronological order. In a number of instances the same statute will relate to more than one area in which case it will be discussed only in the first instance.

## SELECTION OF STAFF

The following statutes relate to the selection of staff:

- Children and Young Persons Act 1933
- Employment of Children Act 1973
- Education (Work Experience) Act 1973
- Disabled Persons (Employment) Act 1944
- Disabled Persons (Registration) Regulations 1945
- Rehabilitation of Offenders Act 1974
- Immigration Rules for Employment of Foreign Workers
- Sex Discrimination Act 1975
- Race Relations Act 1976

## Young People

Children under school leaving age are considered to be employed if they work for an employer even if they are not paid. They should not be employed if under 13 years of age, during school hours if they should be attending school, for more than two hours on any school day or Sunday and on any task where they may be expected to lift, carry or move items that are so heavy as to cause injury.

Allowance is made for a child to undertake work experience programmes approved by the education authority during the last year of compulsory schooling.

Young people, *i.e.* those between school leaving age and eighteen years old, come under the protection of the Factories Act 1961, the Offices, Shops and Railway Premises Act 1963 (OSRPA) and the Young Persons (Employment) Act 1938 (YPEA). YPEA restricts the working hours and overtime worked by young people but this would not affect laboratories that offer a 'standard' working week. The Factories Act makes similar provision and also prohibits the employment of young people on certain processes. Similar restrictions may be imposed by other Acts and Regulations in respect of specific machines.

## MSC Trainees

Young people taking part in TA schemes, such as YTS, are not considered to be employees but have the status of trainees. However, for the purpose of health and safety they are to be treated as being an employee of the organization providing the training (Health and Safety (YTS) Regulations 1983).

## Disabled Employees

Provision is made for the registration of disabled persons under the Disabled Persons (Employment) Act 1944 and the employment of a 'quota' (normally 3%) by every employer of more than twenty people.

## Past Criminal Convictions

The Rehabilitation of Offenders Act 1974 provides that after a certain period of time criminal offences, for which a person has served a sentence, become 'spent' and do not have to be declared to a potential employer.

**Foreign Employees**

*Nationals of European Community States* are allowed to take up employment without first obtaining a work permit. If they wish to stay in the UK for more than six months they will need a resident's permit and the employer will be required to complete the employment certificate on the form.

*Non EC Nationals* must obtain a work permit, for which the employer makes application. Restrictions apply to the issue of such permits and the type of job for which they are issued. Workers eligible include highly qualified technicians with specialized experience and people coming for a limited period of training or work experience.

**Sex Discrimination**

Discrimination on grounds of sex or marital status directly, indirectly, or in the form of victimization, is unlawful under the Sex Discrimination Act 1975. This act also requires that advertisements do not show any intention to discriminate. A detailed Code of Practice provides guidance on the implementation of the correct procedures.

**Racial Discrimination**

The position on racial discrimination is similar to that on sexual discrimination. The Race Relations Act 1976 defines 'racial group' as a person defined by reference to colour, race, nationality, ethnic or national origins. A Code of Practice has also been produced on Race Relations.

**APPOINTMENT**

The main statutes and provisions relating to appointment of staff are:

- Employment Protection (Consolidation) Act 1978
- Equal Pay Act 1970
- Statutory Sick Pay (General) Regulations 1982, as amended by the Statutory Sick Pay (General) Amendment Regulations 1986
- Collective Agreements

**Contract of Employment**

It is not necessary in law for the contract of employment to be in writing although the Employment Protection (Consolidation) Act 1978 requires that each employee be issued with a written statement of employment. This statement should be issued within thirteen weeks of the start of employment. Part-time staff working for less than 16 hours per week are not required to be issued with a statement unless they have been employed continuously for five years or more and work at least 8 hours per week.

The statement must include, or provide notice of, details of:

(i) the parties to the contract, date of commencement with the date when continuous employment began, and job title,
(ii) the scale of renumeration, payment intervals and methods of calculating payment,
(iii) the hours of work, holidays, sick pay and pension conditions (except state pension),
(iv) disciplinary and complaints procedures with specification of the person to whom the employee can apply if dissatisfied, (this does not include health and safety matters).

Alternatively a full written contract may be provided containing express terms (those to which both parties have specifically agreed) relating to the particulars mentioned above. Normally technicians are not issued with such contracts, and many rights and obligations will remain unspecified. These are considered to be implied terms. The principal implied terms are shown in Figure 13.1.

| **Employer** | **Employee** |
|---|---|
| To behave in accordance with good industrial practice. | Fidelity – to serve faithfully. |
| To give reasonable notice if no specific terms have been agreed. | Obedience in respect to reasonable instructions. |
| | To work with diligence, care and honesty. |

**Figure 13.1** *Contract of employment. Common implied terms.*

## Probationary Employees

Although it is common practice for an employer to engage new staff for an initial probationary period this does not entitle the employer to dispense with the services of an unsatisfactory new employee without due notice. The use of the probationary term has little effect on the normal employer/employee relationship.

The Employment Appeal Tribunal laid down guidelines in 1977 for the dismissal of an employee during the probationary period. They expect the employer to show that they have taken reasonable steps to maintain appraisal of the probationer and to give advice and warning.

## Temporary Workers

Such employees may work for a period of time sufficient in length to present a complaint of unfair dismissal. If it has been made clear at the time of appointment that the employee is engaged on a temporary basis the employer's failure to renew the current contract should be considered fair. It is useful to note that if a fixed term contract expires and is then renewed, then dismissal has not taken place.

## Equal Pay Act 1970

The Equal Pay Act implies the inclusion of a clause in every woman's contract in any situation where the woman is doing like, or equivalent work done by a man, to provide for 'equality'.

Article 119 of the Treaty of Rome also provides for equal pay for equal work for men and women.

## Payment of Wages

Until January 1988, the Truck Acts 1983 and the Payment of Wages Act 1960 required that the payment of wages to manual workers had to be in the form of coin of the realm unless they request otherwise. This did not apply to workers such as foreman or technicians.

From 1st January 1987, deductions from wages have been governed by the Wages Act 1986. All employees should receive an itemized pay statement containing the following particulars:

  (i)  gross amount of salary or wages,
 (ii)  details of deductions,
(iii)  the net amount payable, and

(iv) if there are different methods of payment, the amount paid by each.

## Sick Pay

There may be a contractual provision for sick pay and where these exist they will normally run for a specified period of time. From April 1986 the entitlement has normally been 28 weeks in any 3 years. Details of the conditions attached to such payments should be set out in the terms of employment.

Statutory Sick Pay (SSP): since April 1983 all employees are entitled to SSP from their employer for up to eight weeks in the tax year. To qualify the employee must comply with the following conditions:

(1) Be too ill to work for more than four consecutive days (including Sundays and Bank Holidays)
(2) Notify his absence to the employer
(3) Provide evidence of incapacity, normally:
   a self-certificate for 4 to 7 days illness
   a doctors certificate for periods after the seventh day.

If two or more periods of sickness (Periods of Incapacity for Work- 'PIW') are separated by fourteen days or less they are considered to be linked and are counted as one PIW.

Statutory Sick Pay is not payable for the first three days of any period nor after eight weeks during the year. Once the maximum is reached the employee will need to claim state sickness benefit.

## Collective Agreements

In many industries, including the public sector, the payment rates and conditions of employment are determined by collective agreements negotiable between trades unions and employers. The Chapter on Industrial Relations provides more information on employer/trade union relations.

## SAFETY AND WELFARE

The main statutes relating to the safety and welfare of employees are:

- Industrial Training Act 1982
- Health and Safety at Work *etc*. Act 1974

- Factories Act 1961
- Offices, Shops and Railway Premises Act 1963
- Employment Protection (Consolidation) Act 1978

**Industrial Training Boards**

The Industrial Training Act 1964, now replaced by the 1982 Act, established Industrial Training Boards (ITB) in a number of industries employing technicians. The Boards exist to encourage training within the industries under their aegis. They provide services for employers but the employer has no obligation in respect of training other than to ensure that staff do not carry out skilled work without adequate training.

**Safety**

As mentioned earlier in this chapter the employer has a Common Law duty of care under which he should take such steps as are reasonably necessary to ensure the safety of his employees.

Statutory provision for safety is made by the Factories Act 1961 and the Office Shops and Railway Premises Act 1963 in respect of specific items under both the Acts and subsequent regulations. Failure to comply with the requirements may lead to both civil and criminal prosecutions.

**The Health and Safety at Work *etc.* Act 1974**

This Act applies to most areas of employment and lays down general principles supplemented by Regulations and Codes of Practice.

The Act established the Health and Safety Commission (HSC) the Health and Safety Executive (HSE), and provides inspectors with the power to issue improvement and prohibition notices. As with the Factories Act (which it enforces) the HSWA imposes both civil and criminal liabilities. Further details are provided in the chapter dealing with safety.

**Maternity Rights**

The Employment Protection Act 1975, the Employment Protection (Consolidation) Act 1978 and the Employment Act 1980 created the rights for women shown below:

(i) protection from dismissal by reason of pregnancy,
(ii) maternity pay,
(iii) the right to return to work,
(iv) paid time-off to attend antenatal care, and
(v) compensation for unfair dismissal as a result of pregnancy.

Even if a woman is unable to carry out her normal duties as a result of her pregnancy (*e.g.* work with radioactive materials) the employer should offer a suitable available vacancy.

**Maternity Pay**

A woman absent from work due to pregnancy or confinement is entitled to maternity pay. Confinement is defined as the birth of a child, whether living or dead, after 28 weeks of pregnancy. Such pay is made for a period not exceeding six weeks.

*Qualifying Requirements.* The woman must:
(i) be employed until the eleventh week before the expected confinement whether or not she is actually at work;
(ii) have been employed for a continuous period of not less than two years; and
(iii) inform the employer of her intention to take maternity leave at least 21 days before her absence begins, and if intending to, that she wishes to return to work.

*Amount of Maternity Pay.* The amount of pay is calculated at 9/10ths of the week's pay reduced by the Maternity Allowance paid under the Social Security Act 1975.

*The Right to Return to Work.* The woman is entitled to return to work at any time before the end of the period of twenty nine weeks. She should be offered her original employment or if by reason of redundancy her original post has been lost, she is entitled to a suitable vacancy on conditions not substantially less favourable than her original employment.

The employer may request notification in writing that the woman intends to return to work. The actual return to work may be postponed by a date of not more than four weeks by the employer or employee (if supported by a medical certificate in the case of the latter).

**Medical Suspension Payments**

If a factory is closed down in compliance with regulations relating to health and safety, or if an employee is suspended from work on medical grounds in accordance with any regulation concerning safety, they are entitled to renumeration of pay for a period not exceeding twenty-six weeks. An employee is not entitled to such payments if alternative work has been offered and refused.

**Social Security**

There are numerous Social Security Acts and statutory instruments, with the DHSS alone producing over a hundred free leaflets. In respect of employment the main topics are:

- National Insurance,
- Unemployment Benefit,
- Sickness and injury benefit, and
- Means tested benefits for employees.

**TIME OFF WORK**

An employee is entitled to time off work for the following reasons:

- Trades Union officials and members for certain activities,
- all employees for certain public duties,
- an employee facing redundancy to look for another job,
- Safety Representatives to perform certain duties,
- a pregnant woman for antenatal care.

**Trade Union Official**

Employees who are officials of recognized trades unions are entitled to take reasonable time off during working hours with pay to carry out duties concerned with industrial relations and to undergo approved training.

**Trades Union Activities**

An employee who is a member of a recognized trades union may take part in certain activities during working hours.

**Public Duties**

Employees have the right to time off to perform their duties as a:

- justice of the peace, member of a local authority,
- member of a statutory tribunal,
- member of a regional health authority, district health authority or health board,
- educational establishment governor or manager,
- member of water authority or river purification board.

In addition, although there is no statutory requirement to provide time off for jury service it is normally accepted that such time should be allowed. Preventing a person from attending court as a juror is both a crime and a contempt of court, and it is unlikely that any employer or supervisor would wish to put themselves in this position.

## DISCIPLINE AND DISMISSAL

Provisions relating to the discipline and dismissal of employees are:

(1) Industrial Relations Code of Practice TULRA 1
(2) ACAS Code of Practice No. 1. Disciplinary practice and procedure (1977)
(3) Employment Protection (Consolidation) Act 1978

### The Industrial Relations Code of Practice

This was originally issued under the now repealed Industrial Relations Act 1972; it remained in force under the Trades Union and Labour Relations Act 1974 (TULRA) and the Employment Protection Act 1975.

Failure to observe the provision of the code of practice is not in itself an offence but is admissible as evidence. The code gives practical guidance for promoting good industrial relations based on two themes; that collective bargaining should be carried out in a constructive manner, and the importance of good human relations between employers and employees. The main areas of the code are shown in Figure 13.2.

## ACAS Code of Practice No. 1
### Disciplinary Procedure and Practice in Employment (1977)

This code provides practical guidance on how to draw up disciplinary rules and their effective operation. The procedures in the chapter of this book dealing with disciplinary matters are based upon the code.

Dismissal by the Employer

Although generally recognized as meaning the terminating of employment by the employer the word dismissal has no special meaning in common law.

| | |
|---|---|
| Responsibilities | —management, unions, employees |
| Employment Policies | —use of manpower, recruitment and selection training, payment systems, status and security, working conditions |
| Communication and Consultation | —verbal, written with supervisors, with employee representatives, with employees |
| Collective Bargaining | —units, trade union recognition, collective agreements, employee representation |
| Grievance and Dispute Procedures | —individual, collective |
| Disciplinary Procedures | —now superseded by the ACAS code. |

**Figure 13.2** *The main topics covered by the Industrial Relations Code of Practice* (1972).

## Common Law

The common law provisions include the right of an employer to dismiss on a reasonable term of notice and to dismiss summarily for gross misconduct.

## Notice

The period of notice required for termination must be given on the written statement of employment issued to the employee. Where no period is stated the courts will fix a reasonable period.

## Statutory Minimum Notice

The EPCA provides for a minimum notice (Figure 13.3) by the employer which takes precedence over any shorter period stated in the

| Period of Employment (years) | Period of Notice (weeks) |
|---|---|
| 1 month – 2 years | 1 |
| 2 years – 3 years | 2 |
| 3 years – 4 years | 3 |
| 4 years – 5 years | 4 |
| 5 years – 6 years | 5 |
| 6 years – 7 years | 6 |
| 7 years – 8 years | 7 |
| 8 years – 9 years | 8 |
| 9 years – 10 years | 9 |
| 10 years – 11 years | 10 |
| 11 years – 12 years | 11 |
| 12 or more years | 12 |

**Figure 13.3** *Statutory periods of notice that employers are required to give employees.*

contract. It is however open to the parties to agree longer periods if they so wish.

The employee has a statutory obligation to give one week's notice but the contract normally requires a longer period, often the length of one payment period, for example one month. If the employee fails to give the required notice the employer may claim damages for loss caused but such claims are rare.

**Wrongful Dismissal**

This occurs when an employer gives no, or insufficient, notice unless the employee has been guilty of gross misconduct. The employer may give pay for the period of notice in lieu of notice if he so requires.

The right to damages for wrongful dismissal derive from the common law relating to breach of contract and are separate from the right to claim compensation for unfair dismissal under employment protection legislation.

**Unfair Dismissal**

An employee is considered to be unfairly dismissed within the terms of the Employment Protection (Consolidation) Act if:

● the contract is terminated with or without notice, or
● the contract is for a fixed term and that term expires without being renewed under the same contract, or

- the employee terminates the contract in circumstances in which he is entitled to do so because of his employer's repudiatory breach of contract.

For a dismissal to be fair it must be based upon the capability or conduct of the employee or on redundancy.

## Summary Dismissal

The law allows for summary termination of the contract if the employee behaves in a way that is incompatible with faithful discharge of his duty. Such grounds may be theft of the employer's property, drunkenness, or failure to obey a legal order.

## Redundancy

The Redundancy Payments Act 1965 was amended by the 1975 Employment Protection Act and re-enacted in the Employment Protection (Consolidation) Act 1978. To qualify for such a payment the individual must fulfil the following conditions:

- have been an employee for two years ending with the relevant date (any period before the person is 18 years of age does not count).

- have been dismissed in accordance with the Act by reason of redundancy. Dismissal is deemed not to have occurred if the employee's contract is renewed, he is re-engaged under a new contract or offered employment by an associated company.

If offered alternative employment the employee has four weeks in which to give the new job a trial before deciding on whether to accept or reject it.

Employees who have been offered employment on the same terms and conditions as their original job, or suitable alternative employment, but have 'unreasonably refused' the offer are excluded from redundancy payment.

## Fixed Term Contracts

Employees on fixed term contracts entered into after December 1965 will not be entitled to redundancy payment if they have agreed in writing to exclude the right to such payments. Certain employers now include such clauses in all of their fixed term contracts.

The amount of redundancy is based upon the age of the employee, length of continuous service and gross average wage.

## TRADES UNIONS AND TRADE DISPUTES

The main statutes and codes relating to Trades Unions and industrial disputes are:

(1) Trades Union and Labour Relations Act 1974 (TULRA)
(2) Employment Protection Act 1975
(3) ACAS Code No. 2. Disclosure of information to trades unions for collective bargaining purposes 1977
(4) Health and Safety at Work *etc*. Act 1974
(5) Trade Union Act 1984

The legal position of Trades Unions has changed considerably in recent years with the 1972 Industrial Relations Act being repealed by the Trades Union and Labour Relations Act 1974 which was itself amended by the Employment Protection Acts 1980 and 1982.

The 1982 Act has removed the immunity of trades unions from proceedings arising out of strikes or industrial action other than when the action is taken in contemplation or furtherance of a trade dispute. It does not provide immunity for secondary action.

### Recognition of Trades Unions

An employer may formally recognize a trades union or recognition may be implied by consultation with the union on matters such as terms and conditions, allocation of work *etc*.

An employee has certain rights in respect of union membership in relation to the employer as shown below:

- protection from dismissal for union membership or activities—such dismissal is automatically declared unfair,

- remedy in cases of action short of dismissal for union membership or activities, and

- time off work to take part in union activities.

### Disclosure of Information

The Employment Protection Act 1975 provides for the disclosure of information by the employer to the trade union for collective bargaining purposes. The information must be such that the lack of it would impede the bargaining process and that the disclosure of it would be in accordance with good industrial relations practice. Information relating to certain topics, including specifically to an individual who has no consented to its disclosure, is excluded.

The ACAS code of practice gives the following examples of information that could be considered relevant to collective bargaining.

*Pay and Benefits*: structures, grading systems, earnings analysed by group, set *etc*. fringe benefits.

*Conditions of Service*: recruitment policy, training welfare.

*Manpower*: numbers—analysed by, for example, grade, turnover, overtime worked.

*Performance*: productivity data—sales return on investment.

*Financial*: costs, profits, sources of earnings liabilities *etc*.

### Safety Information

The HSAW Act makes provision for the information on safety matters to be provided to trades union safety representatives and for safety inspections and investigations of the workplace. These are discussed in more detail in Chapter 12.

### RETIREMENT

The following statutes relate to retirement:

(1) Employment Protection (Consolidation) Act
(2) Employment Act 1982

The usual date for retirement, in the absence of a specifically agreed date, is 65 for a man and 60 for a woman. Retirement ages do in fact differ from these state pensionable ages with some employers providing for retirement at an earlier age and others allowing the staff to work until a later date.

The employee is precluded from bringing a claim for unfair dismissal after he has reached the state pensionable age or reached the normal retirement age of the organization. Nor is an employee entitled to redundancy payment once the state pension age has been reached.

# The Supervisor and New Technology

## INTRODUCTION

New technology has always been with us and concerned discussion on its impact must have predated the first industrial revolution in the nineteenth century.

In the 1950s–60s the concern was mechanization, the replacement of man's manipulative effort by machine, followed by worries as to the effect of automation, and the replacement of the human control factor by electronics.

In this chapter, the specific applications of computers in research, stock control, or even the 'automated' laboratory will not be considered, but rather the possible effects of the impact of information technology on the role of the supervisor in relation to the organization and as group leader. Some of the implications of two less obvious changes which may have far reaching effects on some areas of laboratory work will be studied; the use of:

- portable and 'side-ward' equipment, *e.g.* that capable of operation outside the laboratory, and
- the replacement of bench tests requiring a moderate degree of technical expertise by simple commercially available test kits.

### Effects of New Technology

The use of microelectronics has been described as the start of the third industrial revolution. The introduction of this new technology has not in itself been revolutionary, just a continuation of the trend from automation; the revolution will be in the effects that the application of microprocessors will have on the workplace and the economy.

In industry these effects have started to be seen, but although widely used in research equipment the new technology has not, to any significant degree, had any impact on the organizational structure of many public sector laboratories. The difference is due to the attitude of management. In industry, information technology is usually introduced as a result of deliberate formal policy discussions, whereas in educational and NHS laboratories, the introduction of computers has often been made on an 'informal' basis by heads of departments, supervisors or research workers.

Equipment purchased following an informal decision is frequently:

- paid for out of the departmental or research budget,
- chosen on the basis of hardware cost and immediate requirements, rather than on a detailed analysis of future needs, or long term performance,
- ordered without consultation with other sections, leading to establishments acquiring incompatible systems often with few application programmes or little capacity for future expansion.

Under these conditions it is not surprising that computers have made little impact on organizational structures of many laboratories.

Logically, as systems and applications proliferate the need for an effective management strategy is increased. Figure 14.1 illustrates some of the objectives for laboratory information systems.

(1)  Hardware and applications software capable of expansion with accessibility to software from other manufacturers or suppliers.

(2)  Ability to communicate with hardware already in use within the organization.

(3)  Reliability of hard- and software.

(4)  Integration of micro- and minicomputer applications software for ease of user interactions.

(5)  Ability for the user to manipulate, customise, store and extract data as simply as possible.

**Figure 14.1** *Factors to be considered when introducing laboratory information management.*

There are widely differing opinions as to what the long term effects of this new industrial revolution will be. Some see a considerable reduction in the workforce while others suggest that there will be new job opportunities. In practice, there will probably be both; an overall reduction in the percentage of those of working age in employment coupled with new opportunities for the employed elite.

Stormier (The Third Industrial Revolution in 'The Effects of Modern Technology on Workers', International Metal Workers Federation, 1979) estimated that at the end of thirty years only 10% of the workforce would be needed to produce the nation's material needs, 90% would be employed in handling information, with 50% of these employed in educational occupations.

In respect of technicians, there are currently shortages of mechanical engineering staff able to understand the applications of microprocessors to mechanical engineering at all stages from that of design to the workbench. There is also a similar shortage of electronic staff skilled in the design elements. It is encouraging to note that electronics technicians made the transition from transistors to integrated circuits, to higher degrees of integration and microprocessors with no significant training problems but this change can be seen as a national progression within a discipline. Problems may well occur when other staff *e.g.* biological technicians, are faced with the full impact of Information Technology (IT) on their own subjects.

Within the chemical industry, work done by Bradbury and Russel ('Technology, Change and its Manpower Implications', Chemical and Allied Products Industrial Training Board, 1980) suggests that as a result of IT there will be a need for more engineers and technicians with a decrease in the numbers of semiskilled and unskilled workers.

Hills ('The Future of the Printed Word in The Impact and Implications of the New Communication Technology', Frances Pinter, 1980) suggests that within higher education the trend will be away from the formal education setting towards more home based systems. The start of this trend has been seen with the development of Open University and Open Tech, but its impact on the educational system, and the employment and role of technicians, is yet to be felt.

## INFORMATION TECHNOLOGY

The new technology may be applied to two main areas: those of process control and management information systems. These are not exclusive and changes on one or both may consequently have effects on other areas such as the organizational structure.

### Process Control

The use of Information Technology (IT) as part of the conversion process ('conversion technology') where it is used as part of the equipment or procedure to convert the input to produce the required end product or output.

In process control, microelectronics or equipment based on microelectronics may be used in:

*Substitution.* Where it is used to update or replace an existing piece of equipment, or where the performance will be improved by the use of microelectronics. The rate of the substitution process will be dependent on the price of the new equipment.

*New Applications.* Where the use of microelectronics will allow tasks to be undertaken which were not possible previously.

*Automation.* Where the equipment is introduced to provide improved performance, reliability, quality or lower production or unit costs. In industry, automation has also been introduced as a means of overcoming skill shortages.

## Management Information Services

The use of this new technology to provide management with information quickly and accurately should lead to improved accountability, cost savings, better stock control and an efficient means of analysing results.

Improved management information systems (MIS) should automatically follow upon improved control systems if they are to be used to maximum advantage.

## The Introduction of Information Technology into the Laboratory

The spread of information technology within laboratories has occurred in three identifiable stages and while the process has not made an equal impact on all disciplines there is no reason to doubt that eventually the majority of laboratories will be, to some degree, automated.

*Computer Controlled Instruments.* The first stage in the application of new technology was the introduction of computer controlled instrumentation. These ranged from the laboratory balance or pipette with its own dedicated microprocessors, to complex analytical instrumentation. Such equipment frequently included automated methods of loading and changing samples.

## Data Handling Laboratory Information Management Systems

The second stage was the introduction of the means of handling the data produced by the instrumentation. These laboratory information management systems (LIMS) enabled test or experimental data to be analysed, samples tracked, costings to be made, and details recorded as to the source of samples.

Initially, the computer controlled instruments were not integrated with the LIMS, requiring data to be transferred manually from a terminal or into the appropriate programme to allow the acquisition of scientific information from the instrument. This was soon followed by the introduction of equipment which made it possible for the users to deal with both scientific and management data by integrating the instrument with the LIMS.

The next step was to link the instruments and LIMS to the organization's main management information computer system.

Examples of how such systems are used can be provided by the Perkin Elmer Computer Aided Chemistry and Everyware (TM) Systems. The Computer Aided Chemistry System (Figure 14.2) links computer technology with analytical instrumentation by providing microcomputer data stations at each instrument. This is extended into a complete laboratory IMS by the use of terminals, for logging and obtaining reports, with Sigma 10 and 15 models and a 3,200 Series Megamini (R) Computer. Figure 14.3 illustrates the roles of three levels of staff within a laboratory using such a system.

In the Everyware product strategy, Perkin Elmer take the LIMS beyond the laboratory and link it to the rest of the organization providing management with a complete MIS.

*Robotic Systems.* The 'automation' of the laboratory by means of microprocessor controlled instruments and computer systems has resulted in many analytical processes being automated but only from the stage of sample introduction to the analysis of the final results.

The introduction of flexible robotic systems now makes it possible for consideration to be given to the automation of sample preparation. At present these tasks are performed manually by technicians and involve such procedures as weighing, adding reagents, and preparing standards. While a considerable degree of accuracy and dependability is required from the technicians the tasks themselves do not require a high level of technical expertise. As such these tasks offer scope for automation using flexible robots. Such systems are now available for a

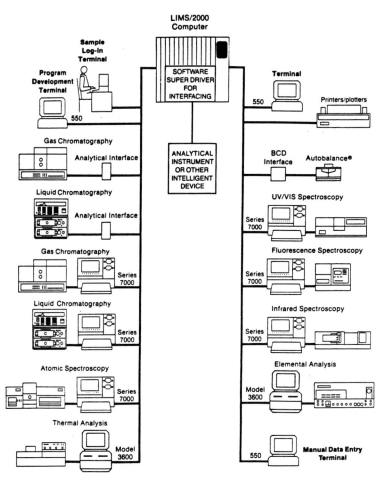

**Figure 14.2** *Perkin Elmer computer aided chemistry laboratory.*
Reproduced from 'Management of Laboratory Information' by S. G. Hurt in
*Analytical Instruments & Computers* (**1**, no. 1).

number of applications and will undoubtedly have an increasing
impact over the next decade.

True robots, flexible automation devices, are defined as multi-
functional reprogrammable devices, and are already capable of a wide
range of accurate movements and dexterities. They have up to five
degrees of movement *e.g.* base rotation, shoulder and elbow joint,
wrist rotation and pitch, and may be made 'intelligent' by means of
sensory input such as cameras and tactile sensors.

| Laboratory Manager | Laboratory Chemist (Supervisor) | Laboratory Technician | LIMS/2000 MODULE |
|---|---|---|---|
| Define systems options<br>Define data sets<br>Terminal assignment | Terminal assignment | | System Configuration |
| Add, modify, delete, list, purge entries<br>Selective reports<br>Define user reports<br>Archive | Add, modify, delete, list purge entries<br>Selective reports<br>Define user reports | | Protected Maintenance |
| Set status | Modify a sample<br>Display a sample<br>Set status<br>Complete a sample | Enter a sample<br>Modify a sample<br>Display a sample<br>Set status<br>Repetitive samples | Sample Management |
| Set status | Complete an analysis<br>Set status<br>Modify an analysis | Complete an analysis<br>Set status<br>Review method | Analysis Management |
| Status summaries<br>Status reports<br>Selective reports<br>Sample analysis reports | Selective reports<br>Status summaries | Work lists | Report Management |
| | Custom programs to access data base | | User Interface |

**Figure 14.3** *User matrix for Perkin Elmer LIMS/2000 application software modules.*
Reproduced from 'Distributed Processing and Laboratory Data Management' by D. P. Binkley and H. W. Major in *American Laboratory* (September 1981).

These flexible devices offer the laboratory supervisor the ability to change the machine from one procedure to another, as against 'hard automation' where the device is restricted to a specific function, *e.g.* an autosampler, and will enable the introduction of robotics into any laboratory where there are a number of repetitive tasks.

## EFFECT OF THE INTRODUCTION OF NEW TECHNOLOGY

### Organizational Effects

Within those organizations with geographically separate establishments the introduction of new technology can lead to greater accountability being delegated to the local management of the individual establishments. However, the opposite is also true; as a result of new technology the individual establishments can be linked to head office allowing increased co-ordination of activities at the local establishments, and providing the means for checking the performance of each unit. A similar situation occurs within the establishment where budgetary and other controls could be delegated to section level.

The balance achieved between these two factors is a matter of managerial choice, assuming management is aware that it has this choice to make. Surprisingly this conflict had not been identified specifically in many organizations adopting new technology (Rothwell and Davidson, 'Technological Change, Company Personnel Policies and Skill Development'. Manpower Services Commission, 1984). It would seem that in most instances, the working arrangements between management, supervisors, and the workers have been reached by trial and error. Obviously decentralization and delegation can lead to improved job satisfaction for those whose jobs are being enhanced. However, this requires a human relations approach to the choice and design of the system. In practice, it would seem that management look to the system rather than to the people. Staff are expected to fit into the system adopted which is often designed to be 'idiot-proof', reducing staff to the status of machine minders.

A slightly different view has been adopted by those considering the effects of computers on management as against systems designed for use or control of processes used by the 'workers' on the factory floor or in an office. McCall and Segrist ('In Pursuit of the Managers Job', Tech: Report No. 14 NC Centre for Creative Leadership, 1980) recorded that at that time, computers had made little impact on management as such, possibly in part because managers see themselves as being leaders, entrepreneurs or communicators. The use of computer systems involving as they do large numbers of specialists and intermediaries may detract from the manager's perception of themselves.

The use of the desk top computer which allows the manager to retain control and privacy may lead to changes in the near future. These changes may not necessarily be to the advantage of supervisors. Woods in the 'Implications of Microelectronics' (Ed. Twiss, Mac-Millan Press, 1981) suggested that it will lead to a trend away from the participative and consultative styles of management towards a more closed system, with the manager dealing more with a VDU than interacting with staff. This would be a retrograde step; the introduction of modern technology should not be seen as an excuse to return to the old authoritarian style of management.

The introduction of new technology may also lead to increased bureaucracy within the organization and a loss of flexibility for members of staff. The greater use of data handling methods, such as stores stock control, may limit some of the informal methods of working that have played an important part in the functioning of the organization in the past. In the example of computerized stores it may not be possible to obtain any goods unless the appropriate requisition forms have been filled in, rather than a casual request over the counter currently used to obtain goods in an emergency.

### Effects on Jobs

There are three main areas, according to Braun and Senker in 'New Technology and Employment' (Manpower Services Commission, 1982), where new technology will have an impact on the content of jobs. While these relate to employment in general they may also be seen to apply to the work of technicians and thus effect the future of the technician's supervisor. The three areas are deskilling, polarization of skills and sophistication.

*Deskilling.* It is argued that there will be an overall deskilling with systems being designed to be as far as is possible 'idiot-proof'. In the interests of increased productivity, or as an unplanned result of updating equipment, more control will pass to machines. In the interests of efficiency the workforce will be required to work to the speed of the machine, leading to what is known as machine pacing.

Control and decision factors may also be undertaken by the

machine. As the equipment becomes more sophisticated and more self-contained the situation could well arise where it will not only require less skilled operatives but may not even need to be situated within the laboratory. Equipment that may be used on-site, in the field or in hospital side wards could have major implications for laboratory staffing levels.

Examples already exist in industrial processes where the quality control laboratory has been superseded by instrumentation using 'in-line' probes measuring a wide variety of parameters. These probes feed data to a computer which in turn uses prescribed criteria to decide on the appropriate actions to be taken. Thus technical requirement has been transferred from a skilled analytical technician to an instrumentation engineer. Where the instrument is designed to take modular components, the expertise of the engineer may be limited to the diagnosis of faults and component replacement.

Farr ('Changing Patterns in the Medical Laboratory'. *Medical Laboratory Sciences* **42**, 1985) considers another aspect of the problems that might arise with the introduction of new technology into laboratories, namely the end of the present discrete 'disciplines' representing specific areas of expertise and practical demand. He argues that clinical laboratory practice is rapidly becoming more scientific and less technical and that the medical laboratory workers of the future will need to be recruited from generally trained scientists. It is very likely, however, that such workers will need to be assisted by a number of laboratory aides, operating at a lower level than the conventional 'technician' or medical scientific officer grade, responsible for the routine tasks associated with new equipment. The use of robotics may even mean that within a decade relatively few staff at this 'assistant' level will be required.

Adapting the organizational structure to take best advantage of the new technology will often present the majority of problems related to the introduction of IT. These problems will be associated with many of the areas discussed earlier in this book, ranging from organizational needs, intergroup relationships and the vested interests of those concerned. In the clinical laboratory situation mentioned above, the relative future roles of the consultant medical staff (the industrial equivalent of whom would be the professional scientist) the graduate scientific officer (technologist) and the medical laboratory scientific officer (graduate or BTEC qualified technician) will undoubtedly give rise to considerable debate. Transition to a new structure is never easy and the adoption of a system appropriate to the needs of the fully automated laboratory will be complicated by the fact that it is the tasks performed by the most numerous groups within the laboratory, those performing the technical work (be they technicians or scientific officers) which will be most under threat.

A comparison of hospital laboratories in the UK and USA may provide a useful illustration of the possible impact of computerization on laboratory systems.

In the UK, each pathology discipline has its own head of department, usually a medical consultant (pathologist) responsible for the provision of the service, with the day-to-day running of the department devolved to the Senior Chief Medical Scientific Officer and others. The disciplines will be grouped together in a simple Department or Division of Pathology with an elected chairman.

The main production workers in the laboratory are those in the medical scientific officer grades (previously Medical Laboratory Technicians) responsible for conducting the necessary laboratory tests and analysis. Medical Laboratory Scientific Officers are educated to BTec Higher National Certificate Level with senior officers qualified to degree level standard.

The use of automation in haematology, clinical chemistry, and probably in the near future to some microbiological techniques has led to some loss of job satisfaction and concern for the future. Under present working systems Cook ('Medical Laboratory Management in Britain and America', *Medical Technologist*, June 1985) wrote that the MLSO produces between 6,099 and 13,985 items of work per quarter.

Within the American hospitals there is a clear management structure, with a non-medical laboratory manager, normally a chief technologist who has obtained a management qualification. There appears little evidence of conflict between the laboratory manager and pathologist. The other major difference would seem to relate to the number of laboratory aides and assistants at 30% of staff in the USA compared with 9% (aides and ancillaries) or 18% (if phlebotomists are included) in the UK. The American grades are Medical Technologist, a graduate grade with four years full time college training, Medical Technical Assistant with two years college training, and Laboratory Assistant trained in-service for one year.

In the USA, hospital laboratories are normally fully computerized. All laboratories are able to keep comprehensive statistics and base their staffing upon a standard measurement of workload. As a result of computerization the laboratories are able to be organized on a fully cost accountable basis and there is a degree of centralization which if adopted in the UK would drastically change the present laboratory structure within the NHS. A single laboratory in the USA could cover an area currently equivalent to that served in the UK by three or four hospital districts (an area with a population of over 750,000 people); only an emergency service is retained at local level. Under the American system, according to Cox, the work load per technologist/aide rises to between 17,712 and 31,141 items of work per quarter, which is considerably higher than the British figures already stated.

It is open to argument as to whether the use of IT in the American hospitals has led to the improved management structure and increased output, or whether the adoption of new technology was a result of better management structure and cost accounting. In either case the end result is the same, with increased output, cost effectiveness, and an element of deskilling within the laboratory.

*Polarization of Skills.* It is argued that the introduction of IT will lead to a polarization of skills with a decrease in demand for the unskilled as equipment becomes more sophisticated. Coupled with this decrease will be a demand for highly skilled staff capable of designing equipment and dealing with software. Within laboratories a major question will arise as to where the polarization of skills will occur. It is possible that the necessary new skills will be acquired by the present technical grades allowing them to add appropriate software and hardware maintenance skills to those of their present discipline. This will, however, require positive policy decisions on the part of laboratory management in respect of training, and career prospects.

There are two alternative scenarios. The first is that new technology will be seen as requiring skills appropriate to graduate technologists/ research assistants with present non-graduate technicians being relegated to the less skilled jobs.

The second is that a multi-disciplinary team of electronic engineers and specialist staff will be developed to carry out the routine maintenance, repair and reprogramming of modern equipment. Within the NHS the use of direct labour to provide routine maintenance of a wide range of medical equipment, including radiotherapy, patient monitoring, and laboratory equipment showed a saving of about 34% over the costs of commercial contracts. As with the first scenario there is the danger that the involvement of laboratory technicians in the work of the laboratory will be reduced with a subsequent decrease in the role of the technical supervisor.

*Sophistication.* The third hypothesis is that as equipment becomes more sophisticated so will the skills that are required by the staff who operate them. This will necessitate management deciding to adopt a human relations approach to the design of systems so that the operators' jobs retain sufficient complexity or variety to maintain job satisfaction. This is contradicted by the attitude adopted by many industrial organizations which have so far introduced new technology. In this context skills may be considered as different from qualifications. The former relate to the competence in performing certain tasks; education and experience relates to the years of training, experience and examinations obtained by an individual. All three of these may relate to obsolete technology and procedures, but may form a valuable background to the new working situation.

Both the polarization of skills and sophistication of the working tasks require an organized approach to the training of the existing workforce and their supervisors.

## NON-INFORMATION TECHNOLOGY CHANGES

New technology need not be high technology in the form of sophisticated electronic hardware on the laboratory bench. Future staffing levels, grades, and even the categories of staff performing the 'laboratory' tests may be significantly influenced by the use of diagnostic or solid phase chemical systems, simplified test kits available commercially and the use of what were previously laboratory procedures outside of the laboratory *e.g.* in hospital 'side wards' or 'on-site'.

Such changes will have two major effects on the role of the laboratory supervisor and subordinate staff.

*Deskilling.* The first and most obvious effect within the laboratory will be a continuation of the deskilling process discussed earlier. This is illustrated by two examples, the first linked to the automation of laboratory tests, the second based on conventional manually performed procedures.

In the clinical laboratory, particularly in clinical chemistry, prepared reagents and kits are becoming increasingly popular for use with a wide range of instrumentation. The use of such kits offers the advantage of simplicity, convenience and a more uniform product. By freeing the technician from the need to weigh out and prepare reagents they also offer a reduction in labour costs (where technician-time is costed). These have been seen as advantages to the technical supervisor as they have freed technical staff from routine preparatory tasks to enable them to spend more time conducting the actual analysis.

The deskilling implications of such kits have not been immediately apparent to the staff involved as they have been freed from relatively low technology tasks, weighing, preparing solutions *etc.*, to perform high technology operations. The fact remains that the accurate preparation of solutions requires a degree of technical expertise while the more glamorous use of advanced instrumentation is becoming increasingly 'idiot-proof'. It is likely that the time will come when the workload of laboratories can no longer increase to the extent necessary to accommodate those released from routine tasks.

Diagnostic microbiological laboratories are one of the areas of laboratory technology that still remain relatively unautomated. They still depend on microscopy, staining, growing organisms on culture plates and using fermentation reactions to identify micro-organisms. New technology using antibiotics, gas chromatography, and automated microbiology systems based on the genetic composition of the organism may eventually change this, but for the present it provides an example of how 'low technology' methods not requiring instrumentation may also deskill some technical procedures.

One of the key processes in the identification of bacteria was based on their ability to ferment or utilize different sugars, and to produce specific enzymes. These tests were performed using individual bottles of the appropriate media selected, and possibly made up, by the technician. Identification of the organism was based on the use of flow charts and tables indicating the different reactions of each organism. These tests are now conducted using strips of reagents produced commercially with the final result being made by reference to a profile number contained in a book of results based on a data bank held by the manufacturer. As with clinical test kits these tests offer simplicity, consistency, increased accuracy and a reduction in labour costs. Although expertise is still required to interpret the final results, basic technical skills are removed from the laboratory.

In a newer innovation, the basic bacteriological microscopy staining procedure can be replaced by the simple expedient of rubbing the micro-organism on a paper strip and noting the subsequent colour change. As with the other systems mentioned this adds to the efficiency of the laboratory, saving technician time and therefore costs.

These deskilling processes will undoubtedly continue with the result that the numbers of staff required to produce a given output will decrease and the ratio of the grades supervised may change.

*'Side-Ward' and 'On-Site' Testing.* The other significant change could be the removal of what were previously considered to be laboratory procedures from the laboratory. In the National Health Service and American hospitals the tendency towards 'side-ward' testing is already causing some concern amongst laboratory staff. The use of microprocessor based instruments and advances in chemistry and biotechnology has led to the production of a range of compact equipment capable of being used with minimum operator expertise while producing results quickly and economically. These instruments may be operated in rooms attached to hospital wards (side-ward tests), or in the physician's office. This gives rise to the acronym SPOT—Satellite and Physicians' Office Testing. The process is not confined to the medical sector, the use of portable instrumentation will make it increasingly feasible to conduct tests 'on-site', or in the field, thus providing the client with a quicker result and reducing the need for expensive laboratory facilities.

An extreme example of how such systems may remove work from the laboratory is provided by the human pregnancy test. This was once a laboratory procedure requiring a degree of technical expertise. Now tests may be purchased from chemist shops and performed at home by an untrained person.

The use of tests away from the laboratory may not necessarily mean that the laboratory supervisor loses complete control over those conducting the test. Technical staff may retain responsibility for operating the equipment, extending the control of the supervisor outside the laboratory and increasing the need for co-ordination with other groups of workers. Alternatively, where the equipment is operated by other categories of staff *e.g.* nurses, research workers or field engineers, the laboratory supervisor may be responsible for training these people, and for providing a quality control service.

## EFFECTS ON THE SUPERVISORY ROLES

Where information technology has been adopted by the organization for use as part of a management information system, the evidence would suggest that there may be a number of likely effects on supervisors.

### Reduction in the Number of Supervisors

The experience of industry would indicate that the introduction of management information systems has been accompanied by a deliberate policy to reduce the number of indirect workers, such as supervisors and others not directly involved in the production process. Within laboratories where most first line supervisors are involved in working at the bench as skilled technicians this may not present a problem provided the technical supervisor can acquire the appropriate skills. However, where there are senior supervisory grades fulfilling administrative and purely supervisory tasks some rationalization may occur.

### Reduction in the Number of Levels

In addition to a reduction in supervisory numbers, there is frequently a decrease in the number of levels within the organization, producing a 'flatter' structure. This will reduce the opportunity for promotion but will offer the benefits of a flatter structure discussed in Chapter 1.

### Change in Role: Erosion

In some of the cases reported by Rothwell and Davidson in the MSC paper on 'Technological Change' previously mentioned, the supervisors had not been involved in the implementation of the IT system or in the training of staff. This meant that supervisory staff were less expert in the use of the system than those they were meant to be responsible for, and consequently lost status and authority. The reduction in the supervisor's authority also affects their dealings with managers, as the managers may now have all the relevant information available from the computer, whereas previously it would have been provided by the supervisor. The supervisory function may also be eroded by the technology in that it could reduce the areas of supervisory discretion as the new system functions to a pre-ordained programme. In addition, the technology may reduce the need for planning, scheduling, and interpreting results, skills the supervisor may have acquired through years of experience. Other human and organizational skills, such as the use of informal contacts and organizational by-passing to achieve quicker results and to liaise with the different groups or units, may also be rendered unnecessary or be taken over by the computer.

The erosion of the status of the supervisor discussed in Chapters 1 and 12 in relation to specialist managers will also be increased as computer specialists become involved in the detailed planning of work systems.

## Change in Role: Enhancement

Much of the preceding sections of this chapter have dealt with what may have been seen as negative aspects of the introduction of new technology; the deskilling of jobs, reduction in supervisory grades and a decrease in the supervisory function. In fact the use of IT and associated organizational changes can offer the supervisor the opportunity to enhance their status and find a fulfilling role away from the laboratory bench. Unfortunately this may be at the cost of a reduction in the number of supervisory jobs within the organization.

## The Management of Change

Where there has been a reduction in the role of the supervisor as a result of the introduction of IT within industry this has tended to be as a result of a deliberate management policy which included a reduction in the numbers of indirect workers.

This approach to the introduction of new technology is less likely to be encountered within laboratories where computerization is usually introduced 'informally' to improve experimental or testing activities in a unit.

Within such systems there is scope for the technician supervisor to enhance their role through the management of change. There is a natural tendency to resist change, to find a degree of security in the familiar. This feeling is not restricted to supervisory staff and once the supervisor has learnt to live with, and adapt to change, they can take advantage of any hesitation in others.

Where the new technology is introduced at laboratory level it tends to be by purchasing one or two items at a time *e.g.* gas chromatography systems rather than a completely automated chemistry laboratory. This provides the supervisor with the opportunity to obtain an understanding of the equipment, its operation, associated software, and the appropriate action to be taken in case of faults developing while a batch of samples is being run. This latter point could be of considerable importance; other groups of workers/specialists may be involved in servicing the equipment, writing programmes and costing results, but the laboratory supervisor should become the expert on actually achieving those results. In Chapter 2, the means by which the supervisor can extend his influence and achieve recognition was discussed, 'new technology' should provide a means of using those techniques.

Freed from the need to monitor the routine sample preparation and testing that went with the use of non-automated techniques, or chasing progress and paperwork, the supervisor should be able to concentrate on monitoring overall results, improving workflow and on the human relations part of the supervisory function.

The use of an automated laboratory system with an appropriate LMIS could result in an enhanced role for the (fewer) laboratory technicians. Systems can be designed to make them the primary users on a day to day basis rather than solely a 'machine minder'. The supervisor who would be responsible for allocating work, approving results and generally overseeing the function of the laboratory, would need to ensure that basic job design factors, discussed in Chapter 5, were taken into account when choosing an appropriate system. In this the supervisor may once again find the opportunity to play the role of an expert, arguing for a system in terms of 'people factors' rather than just hardware.

In short the introduction of new technology will present a considerable challenge to supervisors, but those skilled in supervisory techniques should be more than capable of meeting that challenge.

# Index